Today's Sounds for Yesterday's Films

Palgrave Studies in Audio-Visual Culture

Series Editor: **K. J. Donnelly**, University of Southampton, UK

The aesthetic union of sound and image has become a cultural dominant. A junction for aesthetics, technology and theorisation, film's relationship with music remains the crucial nexus point of two of the most popular arts and richest cultural industries. Arguably, the most interesting area of culture is the interface of audio and video aspects, and that film is the flagship cultural industry remains the fount and crucible of both industrial developments and critical ideas.

Palgrave Studies in Audio-Visual Culture has an agenda-setting aspiration. By acknowledging that radical technological changes allow for rethinking existing relationships, as well as existing histories and the efficacy of conventional theories, it provides a platform for innovative scholarship pertaining to the audio-visual. While film is the keystone of the audio-visual continuum, the series aims to address blind spots such as video game sound, soundscapes and sound ecology, sound psychology, art installations, sound art, mobile telephony and stealth remote viewing cultures.

Titles include:

Anna Katharina Windisch and Claus Tieber (*editors*)
THE SOUNDS OF SILENT FILMS
New Perspectives on History, Theory and Practice

Danijela Kulezic-Wilson
THE MUSICALITY OF NARRATIVE FILM

Aimee Monaghan
THE VISUAL MUSIC FILM

K. J. Donnelly and Ann-Kristin Wallengren (*editors*)
TODAY'S SOUNDS FOR YESTERDAY'S FILMS
Making Music for Silent Cinema

Palgrave Studies in Audio-Visual Culture
Series Standing Order ISBN 978–1–137–42975–9 hardcover
Series Standing Order ISBN 978-1–137–42976–6 paperback
(*outside North America only*)

You can receive future titles in this series as they are published by placing a standing order. Please contact your bookseller or, in case of difficulty, write to us at the address below with your name and address, the title of the series and one of the ISBNs quoted above.

Customer Services Department, Macmillan Distribution Ltd, Houndmills, Basingstoke, Hampshire RG21 6XS, England

---

Today's Sounds for Yesterday's Films

Making Music for Silent Cinema

Edited by

K. J. Donnelly
University of Southampton, UK

Ann-Kristin Wallengren
Lund University, Sweden

palgrave
macmillan

First published 2016 by
PALGRAVE MACMILLAN

Palgrave Macmillan in the UK is an imprint of Macmillan Publishers Limited, registered in England, company number 785998, of Houndmills, Basingstoke, Hampshire RG21 6XS.

Palgrave Macmillan in the US is a division of St Martin's Press LLC, 175 Fifth Avenue, New York, NY 10010.

Palgrave Macmillan is the global academic imprint of the above companies and has companies and representatives throughout the world.

Palgrave® and Macmillan® are registered trademarks in the United States, the United Kingdom, Europe and other countries.

ISBN 978–1–137–46635–8

This book is printed on paper suitable for recycling and made from fully managed and sustained forest sources. Logging, pulping and manufacturing processes are expected to conform to the environmental regulations of the country of origin.

A catalogue record for this book is available from the British Library.

Library of Congress Cataloging-in-Publication Data
Donnelly, K. J. (Kevin J.)
Today's sounds for yesterday's films: making music for silent cinema / K. J. Donnelly, Ann-Kristin Wallengren.
pages cm
Includes bibliographical references and index.
ISBN 978–1–137–46635–8 (hardback)
1. Silent films – Musical accompaniment. I. Wallengren, Ann-Kristin, 1959– II. Title.
ML2075.D665 2015
781.5′42—dc23 2015027263

Contents

Figures

Contributors

Gillian B. Anderson is a conductor and musicologist. She has restored and reconstructed over forty of the great 'mute' film scores and has conducted them in synchronization with their projection at many important film festivals, universities and performing arts centres with many symphony orchestras, most recently *Modern Times* (Chaplin) with the FVG Orchestra in Udine, *Nosferatu* with the Vancouver Symphony, *Stark Love* (Brown, 1928) with Cinemusica Viva NYU at MoMA and *The Birth of a Nation* with the Opera Orchestra of the Teatro Sao Carlos in Lisbon. Her releases include *Nosferatu* (Murnau, 1921) BMG Classics, *Carmen* (DeMille, 1915) Video Artists International, *Haexan* (Christiansen, 1922), *Pandora's Box* (Pabst, 1928) and *Master of the House* (Dreyer, 1925) Criterion Films. Her books include *Music for Silent Films 1894–1929: A Guide* (1988) and (trans.) Ennio Morricone and Sergio Miceli, *Composing for the Cinema* (2013). http://www.gilliananderson.it.

Emilio Audissino is a postdoctoral researcher at the University of Southampton. He holds a PhD in History of Visual and Performing Arts from the University of Pisa, Italy, and specializes in Hollywood cinema, Italian cinema, stylistic film analysis, film dubbing and film music. He has published journal articles, book chapters and encyclopaedia entries. He is the author of the monograph *John Williams's Film Music: 'Jaws', 'Star Wars', 'Raiders of the Lost Ark' and the Return of the Classical Hollywood Music Style* (2014), the first book-length study in English on the Hollywood composer.

Carolin Beinroth studied Musicology at the Justus-Liebig-University of Giessen, Germany, and at the University of Wisconsin, Milwaukee (USA). In 2012, she finished her degree with a master's thesis on the topic 'Music for Silent Movies: An Investigation of its Contemporary Reception in German Music and Film Magazines'. Currently, she is doing her PhD on music in the early years of the German silent cinema. In the course of her PhD, she is also in charge of the music archive of the Deutsches Filminstitut (DIF), in Frankfurt/Main.

Matti Bye started his career twenty-five years ago composing and playing to silent film, being a 'house musician' at the Cinemathèque at The

Film House in Stockholm. He is a regular performer and composer, solo as well as with his ensemble, to various silent film festivals and screenings around the world, for example at Pordenone Silent Film Festival and the San Francisco Silent Film Festival. He has now shifted focus, giving most attention to contemporary film and television productions. He has written awarded scores to, for example, Jan Troell's *Everlasting Moments* from 2008 and *Faro* by Fredrik Edfeldt from 2013.

Beth Carroll is a lecturer at the University of Southampton. Her interests include the relationship between sound and image, the musical, and space. Her monograph *Feeling Audio-Visual Space in Film Musicals* is forthcoming (Palgrave Macmillan).

K. J. Donnelly is Reader in Film at the University of Southampton. He is author of *Occult Aesthetics: Sound and Image Synchronization* (2012), *British Film Music and Film Musicals* (Palgrave Macmillan, 2007), *The Spectre of Sound* (2005) and *Pop Music in British Cinema* (2001); and editor of *Film Music: Critical Approaches* (2001), co-editor (with Phil Hayward) of *Tuning In: Music in Science Fiction Television* (2012) and co-editor (with Will Gibbons and Neil Lerner) of *Music in Video Games: Studying Play* (2014). He edits book series for Palgrave Macmillan and Edinburgh University Press and has recently published *Magical Musical Tour: Rock and Film* (2015).

Michael Hammond is Associate Professor in Film History at the University of Southampton. He has written extensively about silent cinema. He is the author of *The Big Show: British Cinema Culture and the Great War* (2006). His current research is a British Academy funded project concerned with the impact of the Great War on the aesthetic practices of the Hollywood studios between 1919 and 1939. He is also a founding member of the Dodge Brothers who, with silent cinema pianist Neil Brand, play musical accompaniment to silent films throughout the UK and Europe.

Ed Hughes has had compositions commissioned and performed by The Opera Group, London Sinfonietta, *I Fagiolini*, BBC Singers, BBC Symphony Orchestra and many ensembles and soloists. His work has been recorded on two discs for Metier Records and broadcast on BBC Radio 3 and other stations. The New Music Players, an ensemble he founded and directs, has recorded his original soundtracks to films by Sergei Eisenstein and Yasujiro Ozu for release by Tartan and BFI. He won a

2014 British Composer Award for *Chaconne for Jonathan Harvey*. He is currently Head of Music at the University of Sussex.

Matthew Malsky is Associate Provost and Dean of the College, as well as Professor of Music at Clark University. His articles are published in the areas of ethnomusicology, cultural studies and sound/film studies by Wesleyan University, University of Illinois and Bloomsburg Presses, and online through *Reconstructions* and *World Picture Journal*. A volume of his music for string quartet is available through Centaur Records. He has recently released a compact disc of chamber music entitled *Geographies and Geometries* with the Worcester Chamber Music Society, Radius Ensemble, and C-Squared on Ravello Records. His music for silent film may be heard at http://matt.malsky.com.

Christopher Natzén received his PhD in Cinema Studies at the Department of Cinema Studies, Stockholm University in 2010. Since mid-September 2009 he has been a researcher at the Collection and Research Development Division, National Library of Sweden. Currently he is involved in a project to give remote cross-border access to the library's AV-collections. Christopher also works with the ongoing projects filmarkivforskning.se and biografblad.kb.se in order to give open access to cinema related content. He is also co-project leader of the pan-European project EUscreenXL. His research is currently focused on Swedish cinema music culture 1905–1915.

Jeff Smith is a Professor in the Communication Arts Department at the University of Wisconsin–Madison and the author of *The Sounds of Commerce: Marketing Popular Film Music* (1998) and *Film Criticism, the Cold War, and the Blacklist: Reading the Hollywood Reds* (2014). His most recent essays on music in film include contributions to *The Oxford Handbook of Film Music Studies* (2014, ed. David Neumeyer), *Popular Music and the New Auteur: Visionary Filmmakers after MTV* (2013, ed. Arved Ashby) and *Sound: Dialogue, Music, and Effects* (2015, ed. Kathryn Kalinak).

Ann-Kristin Wallengren is Professor of Film Studies, Lund University, Sweden. Her research embraces questions about film and national and cultural identity, representation, ideology and transnationality, as well as on different aspects of film music. She has recently published a monograph in Swedish about the cinematic representation of Swedish-Americans and emigration to the USA, *Welcome Home Mr Swanson:*

Swedish Emigrants and Swedishness in Film (Swedish edn. 2013, English edn. 2014), and together with K. J. Donnelly edited a special issue of *Music and Moving Image* about experimental psychology and film music. Other publications include books and articles about different aspects on Swedish film, film music and educational television.

1

Music and the Resurfacing of Silent Film: A General Introduction

Ann-Kristin Wallengren and K. J. Donnelly

It has become an often-repeated maxim that silent films were not usually silent and that there was habitually some sort of musical/sound accompaniment for them, ranging from the basic (an out-of-tune piano with a limited improvising player) to the elaborate (a large orchestra playing a specially written score). In recent years, there has been something of an explosion in the performance of live music to silent films. Whereas even a decade ago, a silent film with live music was a very occasional treat, these days one does not have to go too far afield to find such events.[1] They appear to offer a unique experience that is qualitatively different from synchronized sound film screenings. This also allows scope for musicians, who are able to improvise and experiment, to write music in a certain idiom, to take the opportunities offered by moving away from the centre of the stage or historically reconstruct the music that would have been heard at the time of the film's release. Among reasons for the burgeoning live music for silent films culture is the elevated cultural status that relatively recently has become associated with silent films. They are generally seen as closer to art, in contrast with the popular forms of cinema evident at multiplexes, and have this potentially high culture status bolstered by their association with cultural heritage and its surrounding discourses. Another reason is the highly evident reduction of live music (such as jazz and rock music in pubs and clubs and in many cases the less public funding available for art musicians to perform – certainly the case in the UK and central eastern Europe), which has made live events more of a rarity and thus more valued.

Silent cinema is more prominent now than it has been since the 1920s. Festivals and archival showing abound, and silent films are part of education, entertainment and art. Indeed, one might argue that the silent film aesthetic has remained alive and well (although less prominent) in avant garde cinema (and evident in films like those of Guy Maddin, for example). In 2012 a silent film in a more conventionally narrative form conquered cinema audiences and for the first time since William Wellman's *Wings* from 1927 a silent film won the Oscar for best picture.[2] We are of course referring to *The Artist*, directed by Michel Hazanavicius, a film which was awarded several Oscars, including for the original soundtrack (which included several prominent references to the highlights of film music history). It is no coincidence that the Academy jury and the world-wide audience were ready to celebrate the silent film art at this time. Around the world we are offered more and more opportunities to experience silent film with music, and a growing number of people have discovered the magic and beauty of silent films with live music.[3] A decade ago it was possible to experience this at festivals in metropolitan centres, such as in London, New York, Paris, Berlin, Los Angeles and San Francisco (and of course at the flagship Pordenone Silent Film Festival). Yet in recent years, silent films with live accompaniment have also resurfaced in smaller nations with different film cultures. To take Sweden as an example, there have been quite limited possibilities to watch silent films: customarily it has been at the cinematheque screenings in Stockholm where a pianist or a small orchestra (usually Matti Bye) has accompanied the films. During the celebration of the anniversary of cinema in the middle of the 1990s, silent film was celebrated in the city of Malmö in southern Sweden with screenings accompanied by symphony orchestras, but this was at the time a very unusual happening in Sweden. Now, twenty years later, even small film festivals in Sweden show silent films, and in the autumn of 2014 the cinematheque in Stockholm commissioned well-known Swedish artists from different music genres to write and perform music to some of the masterpieces of silent film history. The programme committee explicitly asked the question, which is also one of central issues in this book: what will happen when the silent film era meets the 2010s and popular musicians? Further to this, the book asks: with silent cinema so remote, how are current silent film accompaniments conceived?

Where there are silent films, there are musical accompaniments. Many DVD releases have new, specially-written music for the films while others go out of their way to have an 'authentic' score from the time of the film's release. A notable demarcation exists between scores for silent films that espouse the notion of historical veracity and those that aim to furnish something new. At heart, this betrays a profound difference in conception of silent films, on the one hand seeing them primarily as historical documents, works of art essentially imbued with the period of their production, and on the other seeing them as living objects that have new life breathed into them by the performance situation and the new moment of the film's experience with new music. The former approach represents the important current notion of the museum while the latter represents the possibility of reappropriation, which can produce startling novelty but in some cases involve the crude co-opting of existing culture into something of questionable value.

While in the overwhelming majority of cases there is no aim at an incongruous experience, in some cases there is an evident irony of antiquated visuals allied to some very modern sounding music. However, there is rarely an aim at the contradictory. There is almost always a strong sense of unity between sound and visuals on an abstract level perhaps as much as on an emotional and narrative level. There is often a concern with dynamics and the rethinking of cinema space as partially illusory and partially present (the flat space of the film and the actual space of musical performance). In the past, silent films have attracted sometimes improvising pianists and organists, or composers working with orchestras or small art chamber ensembles. In recent years, however, a whole new breed of musician has become involved, not only pop musicians using electronics but also experimental musicians and 'turntablist' DJs, who can spin discs to the action. There are a number of possible reasons for the increased interest of popular musicians, not least the 'crisis' in the music industry and the proliferation of ageing musicians wishing to branch out. There is a massive range of films with live and new scores, that run from the historically accurate orchestral scores by scholar conductor Gillian B. Anderson (for example *Nosferatu* [1922, F. W. Murnau], *Häxan: Witchcraft through the Ages* [1922, Benjamin Christensen] and *Way Down East* [1920, D. W. Griffith]), conductor and composer Carl Davis who composes new scores in a film musical classical manner (for example the Harold Lloyd film *Safety Last* [1923, Fred Newmeyer], *Ben Hur: A Tale of the Christ* [1925, Fred Niblo] and *Show People* [1928, King Vidor]), to contemporary sounds such as pop producer Giorgio Moroder's version of *Metropolis* (original 1927; new version released in the cinema in 1984), which fitted Fritz Lang's film up with contemporary pop songs and electronic keyboard music, while two decades later dance music DJ and mixer Jeff Mills also provided an electronic score for the film. Gothic heavy metal band Type O Negative's music from different albums has been used to create a score for *Nosferatu*. Films such as Eisenstein's *Battleship Potemkin* from 1925 have had new scores furnished by top pop act the Pet Shop Boys and avant garde composer Ed Hughes. Similarly, Dziga Vertov's *Man With a Movie Camera* (1929) is available on three different DVDs with different scores by Michael Nyman, the Alloy Orchestra and In the Nursery.

As this illustrates, DVD release has encouraged the furnishing of new music for silent films. While in some cases, there is a solid attempt to provide unobtrusive music that merges with the historical character of the film, in other cases the intention is otherwise. A recent more avant garde endeavour would be Peter Rehberg and Stephen O'Malley's electronics and guitar drone score (as KTL) for Victor Sjöström's *The Phantom Carriage* from 1921. After live performances, this was released on a DVD known as 'the KTL version' and later their score appeared as an option alongside Matti Bye's score on the high quality Criterion DVD release. Similarly, *Häxan: Witchcraft through the Ages* has also been made available with multiple soundtracks, one DVD including both the William Burroughs version and Geoff Smith's hammer dulcimer accompaniment. Indeed, one highly notable phenomenon of recent years is how single films are made available on DVD with different musical accompaniment, thus making for a different cultural object, if not a different film itself. Arguably, silent films with live musical accompaniment are a completely *new* culture as well as a return to a once-vibrant but seemingly sidelined culture, one of both a different form of cinema aesthetics and a time and place where films retained an immediacy of live performance. In some cases, the rethinking of the film has gone far further. For example, DJ Spooky created a 'remix' of D. W. Griffith's *Birth of a Nation* (1915) as *Rebirth of a Nation*, commenting on the racist assumptions at the heart of the original film in a manner rarely achieved by film historians or cultural critics. Fritz Lang's *Metropolis* was also recast as something approaching a rock opera, with a rock group performing alongside singers and dancers, while the film was screened on two screens at sides of stage. Such situations, where the film ceases to be the central focus are occasionally evident, such as a recent project in New York where an Oscar Micheaux film was screened with live jazz and people reading unrelated poems and prose texts. This kind of performance resembles in a modern way the early stages of silent film music practice when film screenings were part of bigger programme in the cinema theatre or at vaudeville theatres, when the film was only one of different features amongst dances, conjurer's performance, songs, and so on.[4]

The growing interest with the audiences around the world coincides with a swelling number of academic books about silent film music. Research and public interest seems in this way to have cross-fertilized each other, and new kinds of silent film music histories have been produced, inspired by the general movement of a turn to among other things a piecemeal history and attention to local film histories in the so-called historical turn.

This involves a return to the sources and to small nation film culture. For example, in his article 'The Living Nickelodeon' in the book *The Sounds of Early Cinema* Rick Altman strongly advocates a much more differentiated and contextual view on silent film music which implies a wider use of primary material.[5] If scholars writing about silent film music at the end of the twentieth century often pointed out how film music in general and silent film music in particular was unfairly treated, in the twenty-first century we have access to diverse research on the field. Only to mention some of the books from the last fifteen years: the already mentioned *The Sounds of Early Cinema* edited by Richard Abel and Rick Altman, Rick Altman's *Silent Film Sound*, Julie Brown and Annette Davison's anthology *The Sounds of Silents in Britain*, and the recent anthology by Claus Tieber and Anna K. Windisch *The Sounds of Silent Films: New Perspectives on History, Theory and Practice*.[6]

It is no exaggeration to claim that music constitutes a bridge between the old silent film and the modern audience; music is also a channel for non-scholarly audiences to gain an appreciation of silent films. Music has become a means for both musicians and audiences to understand this bygone film art anew. Hence, there is an astonishing parallel to the use of music in the 1910s and the use of music almost exactly a hundred years later. In the 1910s music, as is well known, was an important element in an unofficial campaign to raise the status and cultural prestige of cinema. Novels and theatre plays were adapted, pompous cinema theatres were built, and orchestras or at least chamber ensembles made their entrées. Other types of music, such as opera or romantic music from the nineteenth century, was applied in the cinemas to give a patina of 'high art' to the proceedings. The aim was not simply to make cinema seem more legitimate but also to attract audiences with the promise of something beyond the everyday. Around this time, a different way of connecting music and film developed which resulted in original scores or cue sheets – 'cinema music' became 'film music' in that music became conceived as united with the film. Today music is again at the centre of spreading an interest in silent films, and also is furnishing a different value to this form of cinema, erasing the popular misconception of silent cinema as a primitive form of film which was simply waiting around for recorded sound to come along.

There was undeniably a certain poetic and magic quality connected with the silent film that many artists thought was lost when sound film took over. A substantial portion of these qualities was part of the films themselves, but not least it is imbedded in the music that was played live in the theatre alongside the projected film on screen. Now this magic has transferred to modern media culture with music playing a highly prominent part. This burgeoning culture has so far evaded even the most basic description by scholars let alone any sort of theorization. This book will be the first of its kind in that it aims to bring together writings and interviews to delineate the culture of providing music for silent films. It will include a focus upon scholars whose music is historically accurate and at the other end of the spectrum it will include popular musicians who have no regard for any reconstruction of the film as a historical object but wish to forge it anew as an aesthetic object in the present. It will not only have the character of a scholarly work but also will have something of the manual about how to make music for silent films, in that practical concerns will be addressed. The aim is to approach the subject from more than one angle. Rather than simply looking at music for silent films as a matter for film and music history and historians, it should also be conceived from the point of view of more recent musical practices (both popular and experimental) and art multi-media activities.

The following chapter is K. J. Donnelly's discussion of novel musical approaches to scoring silent film, titled 'How Far Can Too Far Go?' This complements some of the discussion in this introduction and sets out the difference between 'historical' scores (which may be reconstructions or historically sensitive) and 'novel' approaches. This is a companion introduction and deals with the 'big picture' of the developing culture of silent films with live music and general attitudes towards history.

Part I, 'Archives and Historical Practices', begins with a chapter by Carolin Beinroth which addresses the use of archives for silent film music, in particular the archive at the Deutsches Filminstitut in Frankfurt. After this, Emilio Audissino's chapter looks into Gottfried Huppertz's score for Fritz Lang's *Metropolis,* detailing how it has been changed and developed over the years. Then, Gillian B. Anderson discusses her reconstruction of the music for Charlie Chaplin's *The Circus,* pointing to the specifics of the situation for those aiming to pull together material that is as close as possible to the music that would have existed at the time of the film's release. Taking a less strict historical view, Michael Hammond then discusses his involvement in constructing a score that is less scholarly and perhaps more generally historically fitting to a cowboy film starring William S. Hart.

Part II, 'Novel Music and New Issues', begins with Jeff Smith's chapter about the Giorgio Moroder version of *Metropolis* released in the 1980s, called 'Bringing a Little Munich Disco to Babelsberg'. This chapter deals with one of the earliest instances of a novel score and which has retained something of a controversial reputation. This is followed by Beth Carroll's 'Soviet Fidelity and the Pet Shop Boys', which addresses the surprising instance of the electropop band performing music to Eisenstein's *Battleship Potemkin*, and includes an interview with the musicians. Christopher Natzén's chapter, 'Multiple Soundtrack Versions on DVD: Scoring Modern City Life and Pastoral Countryside', investigates how different music can change the experience of the same film, experiences that are made available to a wide audience through the publishing of DVDs with multiple soundtracks among which the spectator can choose. This practice can in some distant manner be compared with the reality in the silent film era where different cinema theatres offered different music. Natzén however, also draws on the influence of technological changes in his chapter.

Part III, 'Current Practices and New Traditions', has a chapter called 'Edit's Hand: Music to *The Phantom Carriage*' by premier Swedish silent film accompanist Matti Bye, which discusses his techniques for creating music for Victor Sjöström's classic film. Matt Malsky then discusses the process he went through in composing music for one of the best-known city symphony films, in 'Scoring Ruttman's *Berlin*: Musical Meaning in Historical and Critical Contexts'. Finally Ed Hughes discusses the processes he has followed in writing modern art music for a succession of diverse silent films in 'Silent Film, Live Music and Contemporary Composition'. The book concludes with Ann-Kristin Wallengren's chapter 'To be in Dialogue with the Film: With Neil Brand and Lillian Henley at the Masterclasses at Pordenone Silent Film Festival', which looks into how the celebrated silent film festival has established classes to instruct the silent film accompanists of tomorrow.

This chapter's title refers to 'resurfacing', which can be taken in two ways. Firstly, in that silent films have come back to the surface after years 'underground', if you like. But secondly, that many of the current scores for silent films 'resurface' the film, in the way that a wall or garden might be resurfaced and in doing so be changed significantly. This book addresses the burgeoning and vibrant culture and it is our intention not only to analyse but also to celebrate the culture of silent films with live music, and to encourage its development further.

Notes

1. This is the focus of Martin Loiperdinger's edited collection *Early Cinema Today: The Art of Programming and Live Performance* (New Barnet, Herts.: John Libbey/KINtop, 2011). There is less about music in the book than might be expected. Frank Kessler's chapter, 'Programing and Performing: Early Cinema Today: Strategies and Dispositifs', discusses the importance of offering a different context to make silent films more accessible but focuses on museums and says little about music, and does not see it as a central agent in the process (p. 140).
2. *Wings* was the first film ever to win an Oscar at the first Academy Awards ceremony in 1929.
3. There were a handful of articles published about contemporary music for silent films: Blair Davis, 'Old Films, New Sounds: Screening Silent Cinema with Electronic Music', *Canadian Journal of Film Studies*, vol. 17, no. 2, Autumn 2008, 77–98; Phillip Johnston, 'The Polysynchronous Film Score: Songs for a Contemporary Score for F. W. Murnau's *Faust* (1926)', *Screen Sound: The Australasian Journal of Soundtrack Studies*, no. 3, 2012, 89–105; and Jan Thorp and Eleanor McPhee, 'Alternate Soundtracks: Silent Film Music for Contemporary Audiences', *Screen Sound: The Australasian Journal of Soundtrack Studies*, no. 2, 2010, 64–74.
4. Rick Altman, *Silent Film Sound* (New York: Columbia University Press, 2004); Ann Kristin Wallengren, *En afton på Röda Kvarn: Svensk stumfilm som musikdrama* (*An Evening at Röda Kvarn: Swedish Silent Film as Music Drama*) (Lund University Press. Dissertation, 1998).
5. Richard Abel and Rick Altman (eds.), *The Sounds of Early Cinema* (Bloomington, IN: Indiana University Press, 2001).
6. Julie Brown and Annette Davison (eds.), *The Sounds of the Silents in Britain* (New York: Oxford University Press, 2013); Claus Tieber and Anna K. Windisch (eds.), *The Sounds of Silent Films: New Perspectives on History, Theory and Practice* (Basingstoke: Palgrave Macmillan, 2014).

Bibliography

Abel, Richard and Rick Altman (eds.), *The Sounds of Early Cinema* (Bloomington, IN: Indiana University Press, 2001).

Altman, Rick, *Silent Film Sound* (New York: Columbia University Press, 2004).

Brown, Julie and Annette Davison (eds.), *The Sounds of the Silents in Britain* (New York: Oxford University Press, 2013).

Davis, Blair, 'Old Films, New Sounds: Screening Silent Cinema with Electronic Music', *Canadian Journal of Film Studies*, vol. 17, no. 2, Autumn 2008, 77–98.

Johnston, Phillip, 'The Polysynchronous Film Score: Songs for a Contemporary Score for F. W. Murnau's *Faust* (1926)', *Screen Sound: The Australasian Journal of Soundtrack Studies*, no. 3, 2012, 89–105.

Loiperdinger, Martin (ed.), *Early Cinema Today: The Art of Programming and Live Performance* (New Barnet: John Libbey/KINtop, 2011).

Tieber, Claus and Anna K. Windisch (eds.), *The Sounds of Silent Films: New Perspectives on History, Theory and Practice* (Basingstoke: Palgrave Macmillan, 2014).

Thorp, Jan and Eleanor McPhee, 'Alternate Soundtracks: Silent Film Music for Contemporary Audiences', *Screen Sound: The Australasian Journal of Soundtrack Studies*, no. 2, 2010, 64–74.

Wallengren, Ann Kristin, *En afton på Röda Kvarn: Svensk stumfilm som musikdrama* [*An Evening at Röda Kvarn: Swedish Silent Film as Music Drama*] (Lund: Lund University Press, 1998).

2

How Far Can Too Far Go? Radical Approaches to Silent Film Music

K. J. Donnelly

Currently, silent cinema has become more prominent than at any time since the end of the 1920s, when it was ousted by the cinema of recorded sound. Apart from festivals and special screenings, silent films are a part of education and art.[1] Where there are silent films, there are almost always musical accompaniments. Many DVD and BluRay releases of silent films have new specially-written music for the films while others go out of their way to have 'authentic' scores constructed to be as similar as possible to what accompanied the film at the time of its initial release. There is a profound difference between scores that are founded on such historical principles and those that aim at furnishing something new for the film. While the former offer the museum and history as guarantor of pedigree, the latter offer a potentially new culture although this can mean breathtaking novelty or a crude hybrid.

In recent years, a whole new breed of musician has become involved, not only pop musicians using electronics but also experimental musicians and so-called 'turntablist' DJs, who can spin discs to the action. There are a number of possible reasons for the increased interest of popular musicians, not least the (perpetual) 'crisis' in the music industry and demographics, particularly the proliferation of ageing musicians wishing to branch out. Perhaps silent films have become like Shakespeare plays, which are restaged in versions ranging from 'Elizabethan original' style to radically 'modern' versions. Or perhaps like some art music, which has garnered painstaking 'historical performance' (such as Christopher Hogwood's work with the Academy of Ancient Music) or 'updates' (such as those by jazz pianist Jacques Loussier, or electronic musicians Walter/Wendy Carlos or Isao Tomita). While opinions vary on these 'originals' and 'updates', there was more controversy tied to the electronic 're-imaginings', suggesting that they might be 'going too far' from any original intention. As my chapter title suggests, some of these new musical scores seem to move a significant distance from the film's original musical horizons, and some would say straying too far.

There are some highly esoteric, crass, outré musical accompaniments to silent films, and some of these seemingly have little to do with events on screen. There has been a simplistic assumption that silent films always had 'appropriate' musical accompaniment. In *Silent Film Sound,* Rick Altman has written about how in the first years of the twentieth century, music for silent films was sometimes played by blind musicians.[2] This is less surprising from a musical point of view as blind musicians were not uncommon in this period, and in the first decade of cinema there was often little concept of unified harmony between film and musical accompaniment. This historical fact might serve as an emblem for similar processes now: where some musicians have a certain *indifference* to film.

Old films, new music

Scholars have taken little notice of the current vigorous silent film culture. It has avoided theory and commentary but has thrived in the hothouse of film festivals. In 2010 for instance, the 13th British Silent Film Festival, at the Phoenix Arts Centre, Leicester, included *The Bridal Party* (Rasmus Breinstein, Norway, 1926) with a new score by Halidor Krogh, *Tol'able David* (Henry King, 1921) with a new bluegrass music score by Damien Coldwell (appearing with Nick Pynn on fiddle and Appalachian dulcimer, and Lee Westwood on guitar), and the Louise Brooks film *Beggars of Life* (William Wellman, 1928) with live music by the Dodge Brothers. (See Michael Hammond's chapter in this book.) This was billed as 'The Dodge Brothers performing to *Beggars of Life*', indicating that the film is *not* the primary object but more of an accompaniment to a concert of sorts. Indeed, the Dodge Brothers only rarely perform concerts with films and the event questions the basic tenets of accepted relationships between silent film and musical 'accompaniment'.

Silent films have spawned a wide-ranging and successful industry in recent years, with tours of screenings with live music and international festivals, such as Pordenone in Italy, which embrace a sizeable number of screenings with live music. There is a degree of vertical integration, with films being toured with musicians and then released on DVD. One flourishing ensemble, the Alloy Orchestra, even have their own 'sister company' called Box 5 who restore film prints to be used by them and others. DVD releases have allowed for multiple soundtracks for the same film, sometimes, but not always, stemming from live performances of music to a film screening. Some films seem to attract a large amount of interest from musicians. For instance, *Häxan* (1922) was released with a choice of three scores: the late 1960s recut with W. S. Burroughs with Daniel Humair and Jean-Luc Ponty (usually called *Witchcraft through the Ages*), Geoff Smith's hammer dulcimer music or Bronnt Industries Kapital's electronic music. There are also DVDs available with music by Matti Bye, Gillian B. Anderson and Art Zoyd. Some of these cases illustrate a rethinking of film scores as more malleable things, which sometimes can retain only a loose relationship with the screen. Indeed, isolated scores on DVD can make a sound film into a new experience through promoting the background score at the expense of dialogue and sound effects. Similarly popular is *The Cabinet of Dr. Caligari* (*Das Kabinett des Doktor Caligari*, [1919]), which has had accompaniment from In the Nursery, Timothy Brock, Rainer Viertlböck, dulcimer music by Geoff Smith, a jazz score by Mark Dresser, and Lynne Plowman with the London Mozart Players (2009). Many of these releases were on the Tartan label, which since has been taken over. The culture is a live as well as recorded phenomenon, with many making successful careers from writing, performing and conducting music for silent film, including: Ben Model (resident piano accompanist at the Museum of Modern Art in New York), the Mont Alto Picture Orchestra, Donald Sosin, Gillian B. Anderson (conductor and composer), Dennis James (Wurlitzer organist), Neil Brand (pianist), Stephen Horne, Philip Carli, Günter Buchwald (composer and conductor), Carl Davis (conductor and composer), Timothy Brock (composer), Robert Israel, Steven Ball, Steven Severin, Jon Mirsalis, the Silent Orchestra, the Alloy Orchestra, and the Paragon Ragtime Orchestra. This list is by no means exhaustive and attests to the extensive and vigorous nature of this 'new' region of old culture.

These silent films on DVD with their alternative musical scores not only expand film into a larger object through the array of accompanying extras, but also align the 'rebirth' of silent film with culture that retains a loose relationship with images, or even almost no direct relationship at all. Silent films might be approached as having a polar approach from their music:

Historically-accurate versions
vs.
'Novel', radical versions

These are not absolute poles but the schema represents more the possibilities of a sliding scale between the two. New accompaniment for old silent films will appear somewhere on the vector between being historically accurate and an attempt at something new and radical. The 'historically-accurate' version retains a traditional character to the score, often guided by knowledge of an 'original version'. The approach is scholarly and the processes often follow those of historical research, with the music as an outcome of archival work. These films are often presented at dedicated silent film festivals attended by aficionados and might be seen as a film-musical manifestation of museum culture. Often they follow the principal sources of music from the period, the film music compendiums of the era, such as Giuseppe Becce and Hans Erdmann's *Kinothek* (a declension of *Kinobibliothek*) (1919) and Ernö Rapée's *Motion Picture Moods for Pianists and Organists* (1924).

The 'novel' versions tend to use a rhetoric of cultural renewal or 'updating' in their aspiration to make the silent film a valued object to a wider contemporary audience. There is a strong tendency for certain types of silent film to be favoured, particularly German Expressionist horror films or other 'popular' silent classics. The films in question are rarely anything too obscure, or anything far beyond the choice of a general audience. *Nosferatu* and *Man with a Movie Camera*, for example, are constant sources exploited by musicians. This augurs a courting of a mainstream audience rather than the aficionados who go to silent film festivals. (One might even suggest these are 'silent films for people who don't like silent films'.) Rhetoric about 'updating' the films abounds. For example, the publicity for the Breakin' Boundaries' event at the Roundhouse in London on 28 May 2008, stated: 'New electronic artists produce their own soundtracks to and reworkings of classic silent movies including *Micropolis* (a short film constructed from *Metropolis*), *Häxan*, an abstracted version of *Seven Seals* and more.'[3] I can find absolutely nothing about a film called *Seven Seals* and wonder if their lack of attentiveness to the films might have led them to mistake Ingmar Bergman's *The Seventh Seal* (1957), a sound film but with large quiet sections.

On the one hand, there is the aim to capture 'original intentions' or 'the author's wishes', approaching film as a *historical* entity, bound by and emanating from its 'own time', and on the other a notion that silent film exists 'to be updated' (perhaps to be 'improved'). According to silent film music conductor and composer Philip Carli: 'An exciting score in an idiom that would be familiar to the original audience for the film leads the modern viewer to accept dramatic conventions that might otherwise seem stilted or even unintentionally comical, rather than highlighting these as more "modern" scores can do.'[4] The contradictory point of view is that modern music allows a point of entry for audiences unfamiliar with the conventions of silent cinema. Whichever approach is taken, music is inclined to function as a *frame* for the film, either enabling us to see as audiences in the past saw the film, or effecting to update the experience and offer us a contemporary point of entry to a difficult and antiquated film.

A further dichotomy that does not quite map directly on to the polar divide under discussion pertains to the aesthetic assumptions of the music supplementing a silent film. This does not involve a simple divide between 'historical' and 'novel' approaches. Generally speaking, images inspire particular sonic strategies and aesthetic approaches. The dominant in cinema has been the notion of a 'homology' of film and music, where aspects such as dynamics, 'mood' and kinetics are treated as if they should 'match'. On the other hand, there is the possibility of merely imposing an existing musical approach or tradition on the film, perhaps even in an indifferent and indiscriminate manner. As a homology music must, to a greater or lesser degree, be subordinated to the requirements of film as image. It is therefore essentially *image-based* (and also dialogue-based in sound cinema). Consequently, since the 1920s the dominant tradition in mainstream cinema has been that after shooting and editing, a film will then have a musical score written to fit its particular requirements. This tradition has established a certain format and style for film scores that marks them out as a distinctive musical form. However, the imposition of an existing musical approach or style yields a different outcome. This approach is arguably closer to the tradition of film musicals, where pre-existing songs or stage shows lead to a certain format, where images become an accompaniment to the primacy of the music. In terms of silent films, this can lead to a situation of a 'gig with images' – and where some scores for silent films are available on CD only (such as accompaniments to *Metropolis* by Art Zoyd and the Club Foot Orchestra), confirming the primacy of the music rather than the film images (or indeed the composite). On occasions, film can appear incidental to the music, following a tradition outside mainstream cinema. For instance, since the 1960s, some clubs have projected almost at random film images which are incidental to the club space and the dominance of music in that space, while avant garde cinema can involve procedures radically different from the cinematic norm. At best, music might be conceived as an equal partner with the silent film, but in some cases it manifests *more* than a partner.[5] Music as an addition is able to remove or add to the original intentions of the film. There are some cases where recorded soundtracks are masked and original intentions of filmmakers dismissed. For instance, avant garde composer Mauricio Kagel wrote a new score for *Un Chien Andalou* (1929) to replace Buñuel and Dali's stipulation to use discs of Wagner's *Liebestod* from *Tristan und Isolde* and a tango. Carl Davis masked Charlie Chaplin's recorded score for 'silent movie' *Modern Times* (1936) in order to play the music live with an orchestra, and Philip Glass wrote music for *Dracula* (1931), which treated the quiet early sound film as if it were a 'mostly' silent film, obliterating the atmospheric silences that dominate the film.

Case study: *Metropolis* (1927, F. Lang, UFA)

As I noted earlier, there are a handful of films that have proven to be perennial favourites for musicians supplying new music for silent films. There have been many musical soundtracks created for Fritz Lang's *Metropolis* (1927) by many different artists, including, but not limited to:

1975 – The BBC version with an electronic score by William Fitzwater and Hugh Davies.

1984 – The Video Yesteryear VHS release contained an original score by Rosa Rio on the Hammond organ.

1984 – The version produced by Giorgio Moroder. This was restored, tinted and had intertitles removed and subtitles superimposed over the images. This was the most complete version at the time, running at eighty minutes. It was compiled and overseen by Moroder, who also supplied an electronic score, as well as writing pop songs for the film which were sung by Pat Benatar, Bonnie Tyler, Jon Anderson, Adam Ant, Cycle V, Loverboy, Billy Squier and Freddie Mercury. The 'Original Soundtrack' remains available on CD and while this version of the film had a cinematic release and a short release on VHS, it has never been made available on DVD. (See Jeff Smith's chapter in this volume.)

1986 – Enno Patalas's scholarly restoration, using newly discovered documentation, and featuring Gottfried Huppertz's musical score from 1926. Huppertz had played the piano during the shooting of

Metropolis as well as providing the first score. This was a leitmotif-based orchestral score, which included the music for the mass of the dead (the 'Dies Irae') at some points as well as fragments from Wagner and Richard Strauss.

1998 – Peter Osborne's electric keyboard score, which often sounds quasi-orchestral, for the JEF/Eureka 139-minute DVD version (released in the UK only). The music has never been made available on CD.

2001 – Jeff Mills's version. The successful DJ provided an electronic score which was used for a few screenings, but has been made available only on CD and never on DVD.

2002 – Art Zoyd's version. The French avant garde band who defy description but owe something to progressive and jazz rock performed music for screenings. The music is available on CD but never has been commercially available as a DVD.

2002 – The digital restoration by archivist Martin Koerber, which included a re-recording of Gottfried Huppertz's original score for the film from 1927. This was the Kino International DVD release, which included more material and lasted 124 minutes.

2008 – There was massive excitement about the discovery of the fullest version yet in Buenos Aires. Martin Koerber restored the film again. The initial screenings in Berlin and Frankfurt used a score performed live by the Alloy Orchestra. The ensemble had been founded in 1981 to play music for *Metropolis*, yet they seemed an anomalous preference as the film was being presented as a historical object and Huppertz's score would have been the clear choice. As if to emphasize this, the later cinematic release with synchronized music and the DVD release used a new re-recording of Huppertz's score. (See Emlio Audissino's chapter in this volume.)

2009 – Jean-Michel Danton's touring stage show called '*Metropolis* Revived', which included live music and lightshows.

This list illustrates the wide range of approaches and artists involved in supplying music for *Metropolis*. Some of these 'versions' of the film have a radically different character from others. In his 'BFI Classic' book on *Metropolis*, Thomas Elsaesser stated:

> The desire to perform *Metropolis*, instead of putting it in a critical or historical perspective, is largely responsible for lending new life to the vision of Lang and von Harbou.... To some, such treatment and especially the New Wave music track compounded heresy with

> blasphemy, adding a special cynicism to iconoclastic insouciance. But it was a bold move... Moroder's score, perhaps by its deliberate anachronisms, induces one to discover Lang's images afresh.[6]

Film scholars are usually aficionados of film and therefore unlikely to espouse such a position. They are more likely to be affronted by the audacity of a disco producer assuming that he can 'improve' a film by one of the century's best directors. This is a brave statement by someone with an investment in film history. In the 1980s, the Moroder version must have sounded brand new and very 'now', but the music dated extremely quickly. It now gives the impression of being older than the more 'timeless' Huppertz version (making *Metropolis* an eighties film rather than a twenties film). Moroder's choice of singers dates the film markedly while his electronic incidental music has worn better. Yet, the very period-specific sound of the songs lends a certain quality to the film that makes it perhaps even stranger. Elsaesser's notion of the sense of modernity regained for the film, now makes the Moroder *Metropolis* appear as a constituent of what is now known as 'retro-modernism', and related cultural subgenres such as eighties postmodernism and 'steampunk', aesthetics that mix the modern and the archaic, but are particularly interested in outdated ideas of modernity.

It was not necessarily alien to ally pop music and silent film. The Moroder version was not 'out of the blue', as there had been an insistent trace of silent film culture that remained around the edges of popular culture, and there was a notable strain that reappeared intermittently in pop and rock music. Examples just before Moroder's version abound: Kraftwerk recorded the song *Metropolis* on *The Man Machine* (1979), and record covers used stills from silent films (there are many examples from just before Moroder's release: Be Bop Deluxe's *Live in the Air Age* [1977] from *Metropolis*, Hugh Cornwell and Robert Williams's *Nosferatu* [1979] from the film of the same name, and Bauhaus's *Bela Lugosi's Dead* [1979] from *The Cabinet of Dr. Caligari*).

A more immediate area where the silent film aesthetic had persisted was in avant garde cinema. Avant garde filmmakers such as Andy Warhol often made films that lacked synchronized sound, consisting solely of images. These usually had some music (live or on disc added when they were screened). Even some of Stan Brakhage's films – which are usually screened in silence – have had live accompaniments recently by Sonic Youth, as have some of Harry Smith's films. While avant garde films could be shown with no sound accompaniment, they often would be accompanied by discs or by live music. Furthermore, there is a marginal tradition in cinema including Mel Brooks's *Silent Movie* (1976), or Guy Maddin's films that use silent or early sound aesthetics (*Careful* [1992], *Dracula: Pages from a Virgin's Diary* [2002] and *My Winnipeg* [2007]), and ranging even as far as pop videos (such as Rob Zombie's *Living Dead Girl* or the Kaiser Chiefs' *I Predict a Riot*). One might even argue that contemporary cinema retains a trace of silent cinema in sequences where there is no dialogue, particularly montage sequences and action sequences, where image track and soundtrack (often dominated by music) take the fore. Indeed, Manvell and Huntley suggested that the extensive music-and-image sequences of *2001: A Space Odyssey* (1968) constituted a return to the modes of silent film.[7] However, it is within the bounds of avant garde cinema that the aesthetic remained the strongest. Indeed, the distance between recorded image and live sound fed the avant garde tradition of *disjunction*, in this case between sound and image, allowing the sound sometimes to perform a role akin to commentary on the film. This is not far from the potential in any silent film accompaniment. Indeed, historian, conductor and composer Gillian B. Anderson noted that during the silent era, live music 'could become an almost living, independent character commenting on the pantomimed drama on screen'.[8] This is, of course, also the case for recent musical accompaniments to silent film. While arguably all soundtracks comment on the films, some make more strident or perplexing comments perhaps than others, and those whose 'fit' is least apparent seem to reveal the most clear audio critique of the visual.

New synergy, new psychology

Perhaps the most out of the ordinary, jarring or disagreeable marriages of film images and new music provide the most comment from music to film. Perhaps they are most critical in the 'gap' between the film images and the added music. Some examples appear to have a significant distinction between sound and image, to 'go too far' in their transposition of the films to new contexts. However, the Pet Shop Boys' score for Eisenstein's *Battleship Potemkin* was a surprise. Instead of their characteristic electropop as accompaniment, they instead mostly wrote music for a fairly-traditional sounding orchestral accompaniment to the film. In fact, there is little about much of the music that gives a clue that it was written by the Pet Shop Boys. (See Beth Carroll's chapter in this volume.)

A more radical event was DJ Spooky's performance/DVD *Rebirth of a Nation* (2005) which was a 'remix' of D. W. Griffith's classic film *Birth of a Nation* (1915) – a film identified with the 'birth of modern cinema' as well as being clearly racist. Spooky not only adds a musical soundtrack but also a voice-over commentary on the film's representations. His remix goes further, removing and repeating sections of the film and using superimposed graphics on the screen, all in aid of a critical discourse on the film that includes the film. Since then, he has gone on to provide the score for Kino Lorber's DVD release of *The Cabinet of Dr. Caligari* in 2015.

A marriage more in the spirit of the director's intentions was KTL's score for Victor Sjöström's *The Phantom Carriage* (1922) which is highly effective but also highly unusual. KTL is a musical duo consisting of American guitarist Stephen O'Malley and British electronic musician Peter Rehberg. Their music is based on droning continuous sound, lacking notable melody or harmonic changes. Its austerity and bleakness is harrowing enough, but allied to Sjöström's horror film, the effect is almost overwhelming. O'Malley is most recognized as a member of SunnO)), the drone avant garde/heavy metal band whose live show aims at sonic effects and is a far remove from concerts of songs by popular musicians.

A similarly innovative move involved British musician Geoff Smith who has produced a number of scores for silent films played alone on hammered dulcimers. These include *Faust* (1926), *The Cabinet of Dr. Caligari* (1919) and *Häxan* (1922), and Smith's music was performed alone on three sets of standing dulcimers, with some singing and electronic looping effects. A review of his performance to a screening of *Häxan* at the Barbican, London, in July 2007 noted that: 'Smith's soundtrack might lack a certain dynamic... but [there is] a lingering musical illusion of drifting in and out of consciousness that effectively mirrors *Häxan*'s dreamlike qualities.'[9] The music is certainly unlike most other music for films, although this relates more to timbre (instrumental sound) than to the way the music functions, as it tries to enhance the moods inherent in the images and the narrative development more than it attempts to impose its own aesthetic on the films.

Vsevolod Pudovkin's *Storm Over Asia – The Heir of Genghis Khan* (1928) was performed at the National Film Theatre, London, on 21 July 2001, with live musical accompaniment by the band of Yat-Kha. Yat-Kha is an acclaimed Tuvan Yenisei throat singer, and his musical ensemble has mixed aspects of traditional music with Western pop and rock. While there is some ethnic veracity to this event, matching Pudovkin's images of Mongols with Mongolian music, the sounds were perhaps too distinctive to not take attention away from the screen.

An event called '*Within Our Gates*: Revisited and Remixed' was staged as part of 'Black History Month' at Ithaca College, New York, in 2004. It was inaugurated by a performance of Fe Nunn's specially-written music for the screening of Oscar Micheaux's *Within Our Gates*. This included live music from a jazz quartet, baroque clarinet solo, African dancing and djembe drumming. It also embraced digital live mixes, including a VJ (video jockey mixing images), and spoken word performances derived from academic, theoretical and analytical analysis of the film. It included contributions from the Body and Soul Ensemble and the Ida B. Wells Spoken Word Ensemble. Anna Siomopoulos and Patricia Zimmermann comment:

> The goal of this project was to rethink the exhibition of politically significant films and to encourage a contemporary audience to engage critically with one particularly important film, *Within Our Gates* (1920). In order to create a new reception context for a groundbreaking silent film, we used live music, digital technology, and spoken word performances; we hoped that this new presentation of the film would provoke audiences to see the cultural continuities and discontinuities between different technologies, and the political implications that these technologies have at different moments of social history.[10]

While this event seems essentially about community, technology is clearly central to the event and, judging by the number of constituent parts, film is merely a component. However, it is the central component around which all else coheres. Siomopoulos and Zimmermann continue:

> the music and words had a dialectical relationship to each other; the music released the images from silence and the past, while the spoken word operated as a distancing device to pull the spectator out of the film and into larger historical, theoretical, and critical debates.... Based on our research into black exhibition practices during the silent period, we decided that the music for our performance would not function subserviently to the filmic text or narrative but as an equal. In other words, we sought to destabilize the film text, reanimate film reception, and complicate film spectatorship through music, spoken word, and multiple voices.[11]

With Ida B. Wells's script spoken by performers *in front* of the film screen, this recontextualization of Micheaux's film allowed for other elements to take to the fore, and despite the prominence given to the spoken word, there was a lot of music, not functioning 'subserviently'.

Less analytical and more plainly spectacular was the JMD Project, Jean-Michel Danton's stage show of '*Metropolis* Revived' in 2009. Its publicity boasted (ambiguously) about updating the film: '"*Metropolis* Revived" is a progressive masterpiece that keeps listeners asking for more, this artist has a more refined controversial approach to the message he portrays through his music' (ReverbNation website). The JMD Project re-situated Lang's film in a multi-media extravaganza, and rather than extending the film, it sounds like it reduced the film to a part of something else. Not only was the event based around multiple screens and light shows but there was, according to the publicity, 'a disco afterwards!' (ReverbNation website). In fact, it seemed to have far more in common with large rock festivals or massive stadium concerts than with cinema screenings as they normally are understood. What the JDM Project underlined is just how far *Metropolis* has ceased to be a film by Fritz Lang produced at a certain historical juncture, and how far it has become an acknowledged part of popular culture, boiled down, reduced, composted into a few important, defining images that summarize particular ways of thinking about things.

Music changes the psychological and emotional landscape of films. Therefore, different music yields a different film. Film historians have written unproblematically about silent films untethered to musical accompaniment as if they were complete unities, consensual objects. A focus on (radically) different scores suggests that they are not. The soundtrack is more than merely a 'bolt-on' to the 'film' but more an essential part of the film experience. Indeed, DVDs with two different soundtracks offer good value as they provide, in effect, two separate films.

Conclusion: audio-visual meditations on history

We might divide these silent film events in a different manner, using broad and non-exclusive categories (films and their music can belong to more than one designation):

(i) Scholarly history
(ii) Empathetic history
(iii) Issue-based history
(iv) 'Populist' history

Scholarly history often gives little concession to contemporary taste and tends to focus on 'authenticity' and fidelity to the historical context of the film's initial release. Scholar-conductors like Gillian B. Anderson have reconstructed scores from historical sources, using primary information and often exploiting the musical compendiums of the period to furnish scores as close to what might have been as possible. In historiographical terms, this is closely related to the processes of traditional academic history, and films screened along these lines usually have a strong sense of 'historical veracity' and scholarly responsibility (as curators of culture). Elsaesser noted 'the phantom "first night" of archivists and historians',[12] and indeed, this notion seems an important guide for this approach.

Coming from a similar point of view, empathetic history is essentially interested in how it must have felt for audiences at the time of the film's release. While historical veracity is often important, it is not as important as the emotional tone of the music and some anachronisms may be acceptable in the service of the film. One might argue that many current pianists, such as Ben Model and Neil Brand, provide music that broadly sounds as if it might have been played at the time although both are happier with emotionally engaging music at the expense of scholarship. This is not to deem their music ahistorical – far from it – but to note that mild compromise of historical veracity is acceptable for the cause of current accessibility. Some have gone further and the rhetoric of many silent film accompanists focuses more on audience emotion than historical authenticity. For instance, Stephen Horne avers that, 'I think that music will play an increasingly important role in presentations of silent films. When I talk to younger audience members, it is usually the "performance" aspect of the event which most excites them. It is as though live music dispels their preconceptions, enabling them to engage with what they are watching.'[13] Emphasis on engagement or immersion in the screening of such films is a very current concern. While this is definitely still a form of history, it has been evident in the wave of conversions of museums into 'experiences', with costumed attendants that act out the past including visitors (such as the short-lived Museum of the Moving Image in London, which was accused of 'dumbing down' complex historical material for children's consumption). In *Shivers Down Your Spine: Cinema, Museums and the Immersive View*, Alison Griffiths discusses interactive and immersive museums as part of the same culture as early and pre-cinema and IMAXes.[14] The concluding section of the book is concerned with the debate about whether museums should be for 'elitist' sober education or 'cheapened' popular spectacle, which sums up one of the central contemporary arguments in dealing with historical objects.

Issue-based history takes historical objects and narrative and imbues them with contemporary concerns. Thus any sense of a proper or 'full' history is replaced by one of partisan approach and acceptance of myriad points of view on the past. This is what Richard Handler and Eric Gable have described in their book on living museum Williamsburg as constructionist 'social history', where the lens of current issues defines the past.[15] In terms of silent films, the '*Within Our Gates*: Revisited and Remixed' event and DJ Spooky's remix of *Birth of a Nation* as *Re-birth of Nation* are excellent instances of the original films being recontextualized by self-conscious current and critical theories and practices. Certainly the latter case embodies the notion of critique, and in cases like these the staging might appear less respectful of the films allied to a sense that the past is problematic (and that audiences are unable to be critical on their own). The addition of words that were not part of the original is often crucial to such 'deconstructions' or expansions. Siomopoulos and Zimmermann noted the dialectical relationship of images and sounds, while the overall *Within Our Gates* event was designed to inculcate a complex response and a critical stance in the audience.[16]

Seemingly taking a fundamentally differing perspective, populist history attempts to reformat and make accessible knowledge about the past to an audience who are assumed to need it structured in a 'modernized' and 'popular' manner. Instances of this abound in contemporary popular discourses of history and museums, although the 'Horrible Histories' series of books are perhaps the most successful example. The Moroder *Metropolis* would have to be the prime illustration of this notion, along with the JMD Project's '*Metropolis* Revived'. The latter appears unproblematically to take *Metropolis* as an object that needs radical updating. Blair Davis describes how he found the addition of modern music allowed students to engage more with the film, as it made the film less alien through its incorporation of music from 'their' culture.[17]

The first two categories (scholarly history and empathetic history) are concerned essentially with history (how it was, how it felt to experience it at the time), while the second two categories (issue-based history and 'populist' history) are more concerned with the 'modern' context of the films. The last two are more interested in the present than the past, in some cases more obviously than others.

While there are many other forms that history can take, music and silent film seems to fall under these rubrics (indeed, in some instances more than one simultaneously). The key is not the images ('the film') but the music added to that film. Music for silent films manifests a 'debate' on the nature of history and objects sourced in the past; it debates the *types* of history. Historiographic process is evident in each film, and indeed they might be seen as part of a 'debate' about how film heritage should be used. Indeed, we might note that there is an undeclared battle for heritage, with more at stake than many might think. Should silent films (and the past more generally) be revered as untouched historical objects or experienced as 'living' culture? Should they be the stuff of glass cases in museums or of 'reformatted' commerce? Culture can take both but will be in trouble should one side of the debate win a conclusive victory over the other. While film (and musical) heritage ought not to allow an 'extinction' of films accompanied in a historical manner, it cannot afford to remove the imaginative and virile culture of film accompaniment as a critical or developmental process.

Notes

1. Rick Altman, *Silent Film Sound* (New York: Columbia University Press, 2004), 4.
2. Rick Altman, 'The Silence of the Silents', *The Musical Quarterly*, vol. 80, no. 4, Winter 1996, 648–718 (648).
3. Advertisement in *The Wire*, 28 May 2008, 87.
4. 'BritishSilents' website. http://www.britishsilents.co.uk, accessed 20 April 2010.
5. Gillian B. Anderson, *Music for Silent Films 1894–1929: A Guide* (Washington, DC: Library of Congress, 1988), xv.
6. Thomas Elsaesser, *Metropolis* (London: BFI, 2000), 58–9.
7. Roger Manvell and John Huntley, *The Technique of Film Music* (London: Focal Press, 1975), 253.
8. Anderson, *Music for Silent Films*, xxvi.
9. Edwin Pouncey, review in *The Wire*, no. 281, July 2007, 66.
10. Anna Siomopoulos and Patricia Zimmermann, 'Silent Film Exhibition and Performative Histories: The *Within Our Gates* Project', *The Moving Image*, vol. 6, no. 2, Fall 2006, 109–11 (110).
11. Ibid., 110–11.
12. Elsaesser, *Metropolis*, 58.
13. BritishSilents website.
14. Alison Griffiths, *Shivers Down Your Spine: Cinema, Museums and the Immersive View* (New York: Columbia University Press, 2008).
15. Richard Handler and Eric Gable, *The New History in an Old Museum: Creating Williamsburg* (Durham, NC: Duke University Press, 1997).
16. Siomopoulos and Zimmermann, 'Silent Film Exhibition and Performative Histories', 110.

17. Blair Davis, 'Old Films, New Sounds: Screening Silent Cinema with Electronic Music', *Canadian Journal of Film Studies*, vol. 17, no. 2, Autumn 2008, 77–98 (92).

Bibliography

Altman, Rick, *Silent Film Sound* (New York: Columbia University Press, 2004).

Altman, Rick, 'The Silence of the Silents', *The Musical Quarterly*, vol. 80, no. 4, Winter 1996, 648–718.

Anderson, Gillian B., *Music for Silent Films 1894–1929: A Guide* (Washington, DC: Library of Congress, 1988).

Anon., 'Out There' concert listings in *The Wire*, no. 292, June 1988, 87.

BritishSilents website: http://www.britishsilents.co.uk, accessed 20 April 2010.

Davis, Blair, 'Old Films, New Sounds: Screening Silent Cinema With Electronic Music', *Canadian Journal of Film Studies*, vol. 17, no. 2, Autumn 2008, 77–98.

Elsaesser, Thomas, *Metropolis* (London: BFI, 2000).

Griffiths, Alison, *Shivers Down Your Spine: Cinema, Museums and the Immersive View* (New York: Columbia University Press, 2008).

Handler, Richard and Eric Gable, *The New History in an Old Museum: Creating Williamsburg* (Durham, NC: Duke University Press, 1997).

Hofmann, Charles, *Sound for Silents* (New York: Drama Book Specialists, 1970).

Manvell, Roger and John Huntley, *The Technique of Film Music* (London: Focal Press, 1975).

Pouncey, Edwin, Review in *The Wire*, no. 281, July 2007, 66.

ReverbNation website. http://www.reverbnation.com/page_object/page_object_bio/artist_744271, accessed 2 June 2010.

Siomopoulos, Anna and Patricia Zimmermann, 'Silent Film Exhibition and Performative Histories: The *Within Our Gates* Project', *The Moving Image*, vol. 6, no. 2, Fall 2006, 109–11.

Part I

Archives and Historical Practices

3

Between Practice and Theory: Silent Film Sound and the Music Archive

Carolin Beinroth

Established in 1949 as the oldest film institute of Germany, the Deutsches Filminstitut – DIF – in Frankfurt am Main, headed by Claudia Dillmann, can look back on a long and eventful history. In 2006 the DIF integrated the former communal Deutsches Filmmuseum including its cinema and collections. Since the founding of the museum in 1984, music for silent films has very often been the centre of attention. Not only did the cinema own a Wurlitzer Theater Pipe Organ (1928), which was used to accompany silent films two to three times a month, but the museum also had organized an exhibition on the topic 'Silent Film Music in Germany' in 1988 as well as the festival 'Silent Film and Music' in cooperation with the Alte Oper Frankfurt and the International Musikzentrum Wien in 1988.

Today, music during the silent film period in Germany is still a topic of interest to the DIF, based on the rich, now merged collections. Not least through the Hollywood production *The Artist*,[1] the silent film and the question of its musical accompaniment has gained more prominence in the public discourse. In recent years, the early *Lichtspiel*[2] in general and also its musical accompaniment are being increasingly discussed in Germany. Participants in these discussions include the media, museums and universities.

Due to its extensive collection of contemporary materials, the music archive of the DIF often serves as the starting point for further research or reconstructions of historic film scores. The music archive, which receives and answers numerous enquiries each year, sees itself as both conservator and innovator. Not only does the DIF support silent film music research by providing original scores, but it also allows the performance or reconstruction of historic silent film scores, respectively, the creation of new film music with the help of historic musical pieces.

A recent example would be the cooperation with the German television station ZDF/ARTE, which reconstructed the movie *Das Cabinet des Dr. Caligari* (*The Cabinet of Dr. Caligari*)[3] and its music. In the following, the different collections of the music archive and their importance and utilization in silent cinema as well as today shall be presented.

Materials in the music archive and their utilization in the past and present

The music department of the DIF belongs to the department 'Collections', which is run by Curator and Head of Collections, Hans-Peter Reichmann. In the course of time, the music archive was managed by many different individuals, among others by current star conductor and musician Frank Strobel. In the past, the music archive incorporated three major fields: historical playback devices, written documents and sound storage media. Today, the music archive comprises two major fields: written or visual documents and sound storage media. These fields, in turn, incorporate different collections that were acquired from private collectors, public figures and through the merging of archives.

Sound storage media

The music archive offers a collection of different sound storage media like Edison cylinders, music rolls, shellac records, records as well as tapes and thus gives insight into the different decades of silent film music right from its beginning. The difficulty of the sound storage collection lies in finding out which of the remaining sound carriers were really used in the silent cinema and which were not. While some of the music rolls for example are labelled 'for pictures', the majority of the sound carriers offer no such explicit proof. The only way to determine which of the recorded songs were actually played as a film accompaniment would be a systematic research of historical print media. Nevertheless, today we can be certain that in general Edison cylinders, music rolls, shellacs and records, to some extent, all played a role in the silent cinema.

The focus of the sound storage media collection is on the early years of silent cinema. Approximately 1,000 Edison cylinders and music rolls are stored in the music archive. Most parts of the music rolls were part of the archive of Mechanical Music and Movie Centre, Cornwall, which was acquired in 1989.

The use of mechanical instruments in cinema was widespread in the USA and in Germany until the end of the silent film era.[4] During the heyday of mechanical music, between 1910 and 1930, many firms specialized in developing certain devices which combined different instruments in one apparatus.[5] Orchestrions became highly developed devices, which could be stopped between songs or switch songs automatically. Still, even though the use of mechanical instruments was widespread, it was criticized starting as early as 1910. As part of the reforms, which started in 1907, music was seen as a means to increase the prestige of the cinema in general. Thus, mechanical music, being a component of travelling cinema performances, was no longer regarded as fitting for the *Ladenkino* (Store Front Cinema). In addition to this critique, it was also mentioned that mechanical music did not fit the pictures and had enormous technical deficits. All in all, it can be noted that mechanical instruments were used in the silent cinema until the 1930s even though they stood in opposition to attempts of reformers, who wanted to win a more upper-class audience for the cinema. Most mechanical instruments played music automatically by reading music rolls (rolls of paper with perforations).

A large part of the music roll collection stored in the music archive consists of rolls produced under the brand name *Automatic Music Roll Company Chicago* by the J. P. Seeburg Piano Company. J. P. Seeburg was one of the largest producers of coin-operated pianos and orchestrions in the USA from 1909 through 1928. The music rolls kept in the music archive were produced for Photoplayers style[6] M, S or R and the label directly states: 'Special for Orchestrions and Motion Picture Players'. Some of the rolls even offer the film's name or at least the genre for which the roll was intended. The rolls mostly feature eight to ten different songs. Other rolls of the collection are from the companies Master Touch Music Roll, and Player Rite Music Rolls. Concerning the music rolls, the music archive's main objective clearly lies in the field of conservation. Even though listening to the rolls is not possible due to the lack of playback devices, the labels offer enough information to allow further research in the field of mechanical instruments and their use in the silent cinema.

Apart from the music roll, the DIF also possesses several hundred Edison cylinders. While the fact that phonographs were present in the early cinema is already established for the USA, it still remains to be proven if the same is valid for Germany. In Germany, the two new devices, the cinematograph and the phonograph, were first introduced in the fairground. Sometimes they were present simultaneously but as contemporary articles show, in most cases they could be found separately. In the *Ladenkino*, phonographs were sometimes established in the entrance area and served as a means to attract customers and not to accompany a movie. The Edison cylinders that can be found in the DIF are mostly *Edison Gold Moulded Records*, which were initially fabricated in the year 1902. The cylinders contain popular American songs and opera pieces. Similar to the music rolls, especially the packaging of the cylinders offers interesting information for further research.

Another highlight of the music archive is the German shellac record series called *Grammophon Cinema*. This series featured instrumental pieces from composers like Johann Strauß and Giuseppe Becce, which were arranged especially for the cinema by Ernst Heinrich Franz Pepping between 1929 and 1930. These records were internationally sold under the label *Polydor Cinema*. The label of these shellacs only states the title of the piece and the name of the composers, leaving the conductor space to write down short notes explaining for which scenes of the film the record should be used.

In addition to the musical illustration of pictures, the gramophone was also used to imitate sounds and thus lend more authenticity to the pictures.[7] So-called *Geräuschplatten* (sound effects records) featured sounds from animals, street noise and train sounds. One of these noise records on display in the DIF was arranged and recorded by Edmund Meisel in 1928. This shellac called *Schlachtenlärm* (battle noises) offered sounds to accompany battle scenes.[8] Another shellac stored in the DIF includes sounds like the ringing of a telephone and different kinds of drumbeats which were used to dramatize the action on screen. The *Grammophon Cinema* series was produced exclusively for the accompaniment of films. The gramophone in general was used to accompany films starting as early as in 1899. As a consequence of the replacement of the wax records by shellacs in 1897, the gramophone could supersede the phonograph once and for all. The gramophone was used due to financial reasons but also often because of the lack of available local musical talent. Especially during the First World War, when a lot of musicians were drafted, the gramophone was the predominant instrument of the silent cinema.

In addition to the collections mentioned above, which offer materials that were used in the silent cinema, the music archive also hosts a huge record collection which gives insights into the history of film music in general but especially of German film music. The record collection largely consists of international film scores dating from the early years of the sound film until the 1990s but there is also a huge amount of compilations of German film *Schlager* (popular songs) of the 1920s. The usage of *Schlager* as an accompaniment for silent films, and later on as part of early sound films, is a very important part of the development of German film music.

Starting during the silent film period, *Schlager* were used frequently for the illustration of a film since the repertoire of the pianists was often dominated by these songs. Nevertheless, this method of accompaniment was criticized by the press. The main point of critique was the overemphasis of a small part of a picture, and also that the songs were used by the musicians as a way to comment on a picture ironically. Still, *Schlager* remained popular and at the end of the 1920s they were firmly established as part of German cinema. The use of records instead of live music became more and more established in cinemas at the end of the 1920s. While in some cinemas the transition to sound film had already happened, others could not afford the expensive new projection devices but nevertheless did not use an orchestra anymore.

Before the release of a new film, many record companies started to assign conductors to record a compilation or even to compose a title song that could be sung by the actor on the screen. This record was sold with the first screening of a movie. By doing this, record companies also recognized the potential of the *Schlager* to increase the popularity of a film and started to sell the most memorable songs separately in record stores. Many compilations which demonstrate the *Schlager* can be seen in the DIF, for example compilations of the *Ufa-Films*.

Another important part of the sound storage media collection is an extensive collection of cinema organ music titled *Cinema Organ Encores*. These records were not made for use in cinema, but they show in retrospect some of the most important organists and their performances in the movie palaces in the USA. The records, which feature recordings of the most popular organists, were mainly recorded in the 1970s.

Moreover, the music archive also owns some tapes that reflect the history of the Filminstitut. Among these are recordings of a reconstruction of the movie *Wagner*,[9] which was shown in the Alten Oper in 1983, and a recording of a performance of the music to *Berlin. Sinfonie der Großstadt* (Berlin: Symphony of a Great City)[10] from 1976.

Written material

In contrast to the media storage collection, the collection of contemporary written material of the silent film era is frequently used. The main part of this section consists of various scores and sheet music. With approximately 1,000 international *Kinotheken* and compilations by different publishers, this collection primarily reflects the use of pre-existing musical pieces as 'illustrations' to accompany film showings. In addition to that, the estates of important protagonists of the silent film era, like Giuseppe Becce and Wolfgang Zeller, offer interesting insights into the composition of original scores of this time.

Materials concerning the musical illustration

The silent film era in Germany was, as in any other country, marked by diversity and heterogeneity. In a time when exhibition conventions for movies were not established yet, the accompaniment strategies also varied a lot. The financial capacity of a theatre owner, the talent of the musicians as well as the location and size of a theatre played a significant role here. The most common musical accompaniment strategy of films throughout the silent period in Germany and the USA was the illustration or compilation. In the absence of any film score or suggestions for illustrations, most musicians of the early cinema played whatever they regarded as fitting for a certain scene. The music for films therefore depended completely on the repertoire and knowledge of the musician. Pianists often played popular tunes, while trained violinists performed classical songs. The selected musical pieces were played one after the other, sometimes with modulation but mostly without.[11] Especially in smaller movie theatres in rural areas, the chosen songs very often did not fit the emotional content of the picture. Also, especially the *Kapellmeister* (conductors) of the countryside were faced with many other problems: the films could not be watched before the exhibition, therefore the *Kapellmeister* had to rely on so-called *Zensurkarten* (censor cards), which only provided the titles of every scene. The *Kapellmeister* now more or less had to guess[12] the content of each scene and choose a fitting title.[13] Accordingly, this kind of music was very often considered insufficient. Even if a compilation was attached to a film by the film industry, there were several problems. Firstly, there was hardly any time for practising and also the owner often did not allow any practising due to financial constraints. Secondly, the *Kapellmeister* was often not able to get hold of the pieces mentioned.[14]

The situation was different in the capitals though. In Berlin, for example, for the first release of a film, a lot of attention was given to the music and the *Kapellmeister* were supplied with tools for support, such as a direct phone line from the conductor's desk to the projection room.[15] Also, there was a library with sheet music for compilations. Popular *Kapellmeister* like Giuseppe Becce, Willy Schmidt-Gentner and Werner Schmidt-Bölcke perfected the compilation system in such a way that it resembled original compositions for films.[16] The sheet music stored at the music archive offers lots of examples of songs that were used for compilation purposes. Not only does it feature the collection of musicologist Bernd Heller, with numerous pieces from opera, operetta and classical songs, arranged for the salon orchestras, but there is also an enormous collection of popular songs and traditional German songs that demonstrates which kinds of songs early pianists and orchestras played in the cinema.

Additionally, there is a huge collection by different publishers who printed compilations of music that was especially arranged for the cinema. Attempts to improve the quality of film music increased in the years 1919 to 1921. Famous composers like Giuseppe Becce began to publish their ideas on how to accompany films and also the major sheet music publishing houses saw a chance to improve their incomes by supplying conductors and musicians with film music.[17] The compilations of the most important publishers, called Mignon Berlin, Apollo Berlin, Iris Recklingshausen, Vogger Stuttgart and Schlesinger'sche Buchhandlung Berlin, are all on display in the music archive. The sheet music books of these publishers feature salon orchestra arrangements containing all kinds of different musical pieces. Some of these publishers primarily focused on a certain genre. H. Becker from Berlin, for example, published collections with Schlager for comedies. Apart from the already mentioned German publishers, the DIF also owns many international collections, for instance the *Bosworth Movie Collection* from England. Some of the publishers also organized film music competitions, seeking to encourage more and more musicians to compose their own film music. Heinrichshofen's Verlag, a publisher from the German city of Magdeburg, initiated a film music competition in 1925. The composers were asked to compose pieces that fitted certain emotional moments like 'thunderstorm', 'love' and 'war'. The winners were to be published in a collection which today is kept in the DIF's archive.

Giuseppe Becce as illustrator

The most important contributor to the development of the compilation practice in Germany was Giuseppe Becce. As part of the attempts to create more fitting music to pictures, Becce started to compose short pieces of music for certain moods. His so-called six *Kinotheken* – in all 82 musical pieces – were scoring segments that could be shortened or prolonged and which applied to diverse film subjects. The scoring segments are arranged in different volumes according to certain themes or emotions. It was first published by the Schlesinger Buchhandlung in

1919. The volumes were, for example, called *Tragisches Drama* (tragic drama), *Großes Drama* (grand drama), *Ernste Intermezzi* (serious interludes) and *Exotisches* (exotic). Collected in these volumes were the thematically fitting musical pieces. As an example: *Kinothek 3a*, called *Großes Drama*, consists of the pieces Erscheinen des Todes (Appearance of Death), Höchste Gefahr (Great Danger), Kampf, Tumult und Brand (Fight, Tumult and Fire), as well as Katastrophales Ereignis (Catastrophe). Therefore, these pieces were intended for dramatic films which needed a dark and dangerous timbre. The utilization of these pieces promised to be a very adaptable way to accompany films and it also spared the conductor from spending several hours arranging a compilation. Becce himself criticized the practice of compilations with pre-existing pieces. According to Becce, 'The illustrator took musical pieces from operas, operettas, symphonies, etc. and patched them up to an accompaniment which was devoid of everything but tastelessness and lack of style.'[18]

The *Kinotheken* were greeted with enthusiasm in Germany at this time, but also today the value of Becce's *Kinotheken* cannot be overestimated. This assumption is verified, for example, by the daily enquiries that reach the music archive. Most of last year's requests, whether made by international researchers, musicians, producers or television networks, concerned Becce's *Kinotheken*. Music researchers mostly aim to get insights into the compilation practices of the silent film era by studying the modular design of the *Kinotheken*. Enquiries by musicians and conductors are usually concentrated on the use of the *Kinotheken* for performances of silent films.

The same can be said about requests by television station representatives. For instance, during the last year, the DIF provided selected pieces of the *Kinotheken* for a reconstruction of the movie *Das Cabinet des Dr. Caligari* (*The Cabinet of Dr. Caligari*)[19] in cooperation with ZDF/ARTE and the Murnau Stiftung. In addition to a screening at the *Berlinale 2014*, where the American jazz musician John Zorn provided the music, there was another music production for DVD. For this production, a film music class from the Conservatory of Freiburg, under the guidance of Professor Cornelis Schwehr, arranged film music based on several of Becce's *Kinotheken*.

In 1920, Becce composed a score for the second première of *Das Cabinet des Dr. Caligari*. This music is lost today, but Becce integrated four pieces of this music, which are known today as Großes Drama, Hochdramatisches, Ernste Intermezzi and Verschiedenes, into his *Kinotheken*.[20]

All in all, it can be said that when it comes to the strategy of compilation, the Filminstitut acts as a preserver as well as an innovator. By providing researchers with contemporary materials, the DIF helps to put together new scores or reconstruct old ones with only a few surviving pieces. Most clearly, this can be observed in the example of the *Kinotheken*. Concerning these, the DIF turned, in a wider sense, also into an innovator by contributing to the reconstruction.

Original compositions

Next to the already mentioned contemporary materials concerning illustrations, the DIF also holds various original film scores of the German silent film era. As a priority, Giuseppe Becce and Wolfgang Zeller should be mentioned here, since their estates are stored in the DIF. However, there are also a few scores by Gottfried Huppertz and Edmund Meisel in the music archive.

Giuseppe Becce

In addition to Becce's *Kinotheken*, his original scores (for 90 films) also are among the most frequently used materials of the music archive. It should be mentioned in this context that these scores, with only a few exceptions, belong to sound films of the years 1931 to 1968. One part of the scores belongs to the category of feature films, namely the genres of mountain film and *Heimatfilm* (regional film/sentimental film). Other scores were composed to accompany documentary films and *Wochenschauen* (newsreel).

The 5,100 pages which make up the legacy of Becce were inventoried in the years 1994 to 1995. Since then, requests for Becce's scores have multiplied. Involved parties are scientists, the media, leading research institutions, musicians and freelance musicologists from all over the world.

Regarding the nature of these enquiries, it can be seen that there is an increasing trend to reconstruct films and their music as authentically as possible. Especially because of this development, access to contemporary materials is indispensable.

Even though the scores that can be found in the DIF are for sound films, one can still learn a lot about Becce's compositional thoughts: many of the scores are only available in fragments, and during their inventory it became apparent that Becce often used parts of his scores not just for one movie but integrated it into several movie scores. Similarly to the *Kinotheken*, Becce's original scores can also be seen as set pieces, which are randomly combinable and extremely versatile. Apparently, the original scores as well served as an archive for Becce, which he reverted to very often. In light of this fact, his way of dealing with original scores seems to be similar to the compilation practice. Enquiries about these compositions very often focus on fragments of scores and reapplication in other scores. Also requested are short sketches which Becce wrote down while composing a score, since they give insights into Becce's composition process.

Many enquiries also concern Becce's scores to the silent films *Der Letzte Mann* (*The Last Laugh*)[21] and *Tartüff* (*Tartuffe*).[22] These scores are especially interesting because they offer explanations regarding the action on the screen and advice regarding the volume and tempo of the music. Many requests have been made particularly concerning *Der Letzte Mann*, with the aim to reconstruct the film as authentically as possible. One request came from the Goethe Institute in Lebanon, which eventually organized a screening of the film with a pianist playing the score.

Next to the scores of Becce, some of his journalistic works are located in the music archive, such as the film music composer magazine *FilmTonKunst*, in which he gives advice concerning various points of interest for film music conductors and musicians.

Wolfgang Zeller

Another very important inheritance, stored in the music archive since the beginning of the 1980s, is the legacy from silent film music composer Wolfgang Zeller, consisting of 138 original scores. The collection was inventoried in 1991.

Wolfgang Zeller, born in 1893, became famous after he composed music for the silent animation movie *Die Abenteuer des Prinzen Achmed* (*The Adventures of Prince Achmed*)[23] by paper-cutting artist Lotte Reiniger. After the successful reception of this film music, Zeller was asked by cinema avant-gardist Walter Ruttmann to compose a score for the film *Melodie der Welt* (*Melody of the World*).[24] Zeller, who was not a member of the Nazi party, still composed in the following years music for many Nazi-influenced propaganda films such as for the anti-Semitic film *Jud Süß*.[25]

In 1998/9 the DIF reconstructed the movie *Die Abenteuer des Prinzen Achmed* and, in its course, also the score for the film, which was obtainable in a complete piano version from the Library of Congress. Since the score also featured small pictures of some scenes, it was a very important element for the reconstruction process and helped to confirm the order of the different scenes. Enquiries concerning the score for *Die Abenteuer des Prinzen Achmed* mainly come from researchers who are interested in the combination of pictures and notes in the sheet music.

In addition, other scores of the Zeller collection are often requested due to questions concerning reconstructions. As examples, the scores for *Luther*[26] as well as *Immensee*[27] should be mentioned here. The latter were both reconstructed by the European Filmphilharmonie.

Further highlights

Apart from the original compositions by Becce and Zeller, the score to *Metropolis*,[28] *Panzerkreuzer Potemkin* (*Battleship Potemkin*)[29] and *Die Nibelungen* (*The Nibelings*)[30] are the most valuable for the music archive. Concerning these scores, the DIF often receives enquiries about live performances and research. Another good example for original compositions for silent films is the score for the film *Der erste Weltkrieg* (*The First World War*)[31] by Marc Roland. This score shows how far composers aimed to support conductors in everyday cinema. Roland describes here, in a foreword, which tempo and which volume should be used for different parts of the score and offers advice on how to most adeptly perform the music. Also, he directly addresses theatre owners and explains that the film and the music are meant to be performed together and that they should not show the movie without the music. Thus, by offering this and similar structured scores for further research, the DIF helps to promote knowledge about the various approaches by different protagonists concerning music in the silent cinema.

Books

Another integral part of the music archive comprises historical books concerning music in the German silent cinema. As an example, the exceptional book *Das Musik-Chronometer und seine Bedeutung für Film-, Theater- und Allgemeine Musikkultur* (*The Music Chronometer and Its Importance for Film, Theater and Music Culture*) written by Carl Robert Blum should be mentioned here.[32] This book, which is extremely rare today, deals with one of the attempts of the 1920s to synchronize music and pictures. During the 1920s, many mechanisms and apparatuses were developed to allow a synchronized interplay of music and picture.

In addition to several books on Fritz Lang the music archive also owns a collection of contemporary newspaper articles which deal with various performances of silent films and newspaper articles on silent film related events that were organized by the institute. Also, this collection contains much important information on the organs of German movie theatres as well as conductors and other important protagonists of German silent cinema.

Further activities of the DIF in the field of silent film sounds

Whereas the music archive is mainly active in the role of conservation, other parts of the DIF, especially the film archive, are directly involved in reconstructing highly valuable materials from the silent film era. In the last year, restorer Anke Mebold reconstructed several *Tonbilder*. *Tonbilder* can be seen as the equivalent to the American tradition of *Illustrated Songs*. As the term *Tonbilder* (*Ton* = Sound, *Bilder* = picture) shows, they were a combination of film and sound. During the silent film era there were many experiments with synchronization devices in order to combine opera sequences with silent films. *Tonbilder* were among these attempts. They were usually films of 3–4 minutes in length but they could also last up to 20 minutes. In contrast to the *Illustrated Songs*, *Tonbilder* were films of a show act, a short appearance on stage or a popular opera scene. Simultaneously with the screening of the film, a phonograph or gramophone played the song which was performed in the picture. The most important difference between the *Illustrated Songs* and the *Tonbilder* therefore is that the *Tonbilder* rarely conveyed any narrative message. Starting in 1907, specific *Tonbildtheater* opened, for example in Frankfurt am Main. These showed exclusively *Tonbilder*, but other non-specialized theatres also included them in their programmes. Here, *Tonbilder* and moving pictures were shown in turn.

Anke Mebold acquired several of the short films that were used to accompany the phonograph and set out to find the music that was once played to accompany them. Sometimes, it was only possible to define the nature of the song by the information visible in the picture (for example: How many people were singing? Male or female? Did the setting look like a comedy or an opera piece?). In other cases, contemporary articles helped to determine which songs were originally played to a certain picture. Later on, Anke Mebold synchronized the music with the picture and in this way reconstructed several *Tonbilder* in exactly the way they were originally shown. The *Tonbilder* were presented in the Wiesbaden Caligari Cinema and at the Giornate del Cinema Muto in

Pordenone, Italy, and the records are now stored in the music archive of the DIF.

Perspectives

As a conclusion, it can be said that throughout the years the DIF has contributed to various projects concerning music in the silent cinema or music for silent films. On the one hand, by providing students and researchers with primary materials, the DIF supports further important scientific research and contributes to keeping the tradition of silent film music alive. On the other hand, the DIF also enables musicians, TV stations, etc. to reconstruct historic film scores by providing contemporary materials.

Covering primary auditory materials like music rolls, Edison cylinders and shellac records as well as historical sheet music, the DIF offers insights into the most important aspects of musical accompaniment in German silent cinema.

Next to the contributions in research and reconstruction, the DIF is also committed to the promotion of film studies in Germany. In order to promote education in film science,[33] the DIF is working in close cooperation with the course 'International Master of Audiovisual and Cinema Studies of the Hessian Academy of Film and Media'. Within this framework, the DIF contributed to the seminar 'Filmmusik-Geschichte, Werkstatt, Analyse' (Film music history, workshop and analysis) in cooperation with Nina Goslar from ZDF/ARTE. As part of the seminar, the movie *Berlin. Die Sinfonie der Großstadt* was shown to the students in the movie theatre of the DIF, with an introduction by composer and musicologist Bernd Thewes, who arranged the music for the film.

Due to increasing interest in silent film and its music in recent years, the DIF aims to provide the general public with access to the materials stored in the music archive and thus contributes to the further examination of a part of German culture that still holds many secrets.

Notes

1. *The Artist*, directed by Michael Hazanavicius, 2011.
2. Outdated term for the medium *film*.
3. *The Cabinet of Dr. Caligari*, directed by Robert Wiene, 2014 (orig. 1920).
4. Author Unknown, 'Die Bedeutung der Musikwerke für den Kinobetrieb', *Der Kinematograph*, no. 655, 1919.

5. Karl-Heinz Dettke, *Kinoorgeln und Kinomusik in Deutschland* (Stuttgart: J. B. Metzler Verlag, 1995), 25.
6. The term 'style' refers to the kind of orchestrion the roll could be played back with.
7. G. Kleibömer, 'Aus dem Reiche der Töne: Kinematograph und Grammophon im Dienste der Kunst', *Der Kinematograph*, no. 127, 1909.
8. Author Unknown, 'Geräuschmusik auch für kleine Kinoorchester ermöglicht', *Das Kino-Orchester*, no. 20, 1928.
9. *Wagner*, directed by Carl Fröhlich/William Wauer, 1913.
10. *Berlin: Symphony of a Great City*, directed by Walther Ruttmann, 1927.
11. Kurt London, *Film Music: A Summary of the Characteristic Features of its History, Aesthetics, Techniques; and Its Possible Developments* (New York: Faber and Faber, 1936), 53.
12. To gain an impression of how little information these cards actually carried, they can be reviewed in the text archive of the DIF.
13. Author Unknown, 'Zensurkarten-Jammer', *Das Kino-Orchester*, no. 11, 1925.
14. Author Unknown, 'Die Kinomusik', *LichtBildBühne*, no. 2, 1919.
15. Volker Spiess, *Stummfilmmusik gestern und heute: Beiträge und Interviews anlässlich eines Symposiums im Kino Arsenal am 9. Juni 1979 in Berlin* (Berlin: Verlag Volker Spiess, 1979), 46.
16. Rainer Fabich, *Musik für den Stummfilm: Analysierende Beschreibung originaler Stummfilmkompositionen* (Frankfurt am Main: Peter Lang, 1993), 20.
17. Konrad Ottenheym, *Film und Musik bis zur Einführung des Tonfilms: Beiträge zu einer Geschichte der Filmmusik* (Berlin: Friedrich-Wilhelm-University, 1944), 55.
18. Becce quoted in Michael Beiche, 'Musik und Film im Deutschen Musikjournalismus der 1920er Jahre', *Archiv für Musikwissenschaft*, no. 63, H.2. (Stuttgart: Franz Steiner Verlag, 2006), 94–119.
19. *The Cabinet of Dr. Caligari.*
20. David Robinson, *Das Cabinet des Dr. Caligari* (Basingstoke: Palgrave Macmillan, 2013), 55.
21. *The Last Laugh*, directed by Wilhelm Friedrich Murnau, 1924.
22. *Tartuffe*, directed by Wilhelm Friedrich Murnau, 1926.
23. *The Adventures of Prince Achmed*, directed by Lotte Reinger, 2005 (orig. 1926).
24. *Melody of the World*, directed by Hans Kysers, 1929.
25. *Jud Süß*, directed by Veit Harlan, 1940.
26. *Luther*, directed by Hans Kysers, 1927.
27. *Immensee*, directed by Veit Harlan, 1943.
28. *Metropolis*, directed by Fritz Lang/Erich Pommer, 1927.
29. *Battleship Potemkin*, directed by S. M. Eisenstein, 1925.
30. *The Nibelungs*, directed by Fritz Lang/Erich Pommer, 1924.
31. *The First World War*, directed by Leo Lasko, 1926/7.
32. Carl Robert Blum, *Das Musik-Chronometer und seine Bedeutung für Film, Theater- und Allgemeine Musikkultur* (Leipzig: F. E. C. Leuckar, 1926).
33. In order to introduce the silent film and its accompaniment to children, the DIF organized a project, headed by Andrea Haller, during which the students made new musical arrangements for historic silent films (the project was organized in cooperation with the Hochschule für Musik und Darstellende

Künste, Frankfurt am Main and the results were later on released on DVD as a means to support high school teachers in introducing the genre of silent film to their students).

Bibliography

Films

Adventures of Prince Achmed, The. Produced and directed by Lotte Reiniger. Germany/Austria/Switzerland: ARTE Edition/Absolut Medien, 2005 (orig. 1926).

Artist, The. Produced and directed by Michael Hazanavicius. USA: Warner Bros., 2011.

Battleship Potemkin. Produced and directed by S. M. Eisenstein. USSR: Mosfilm, 1925.

Berlin: Symphony of a Great City. Produced and directed by Walther Ruttmann. Berlin: Deutsche Vereinsfilm AG, 1927.

Cabinet of Dr. Caligari, The. DVD. Produced and directed by Robert Wiene. Wiesbaden: Friedrich-Wilhelm-Murnau-Stiftung, 2014 (orig. 1920).

First World War, The. Produced and directed by Leo Lasko. Germany: Universum Film AG, 1926/27.

Immensee. Produced and directed by Veit Harlan. Germany: Deutscher Fernsehfunk, 1943.

Jud Süß. Produced and directed by Veit Harlan. Germany: Terra Film, 1940.

Last Laugh, The. Produced and directed by Wilhelm Friedrich Murnau/Erich Pommer. Germany: Union-Film der Universum-Film AG, 1924.

Luther. Produced and directed by Hans Kysers. Germany: Cob., 1927.

Melody of the World. Produced and directed by Hans Kysers. Germany: Cob., 1929.

Metropolis. Produced and directed by Fritz Lang/Erich Pommer. Germany: Union Film der Universum-Film AG, 1927.

Nibelungs, The. Produced and directed by Fritz Lang/Erich Pommer. Germany: Union-Film der Universum-Film AG, 1924.

Tartuffe. Produced and directed by Wilhelm Friedrich Murnau. Germany: Union-Film der Universum-Film AG, 1926.

Wagner. Produced and directed by Carl Fröhlich/William Wauer. Germany: Oskar Messter, 1913.

Historic magazines

Author Unknown, 'Die Bedeutung der Musikwerke für den Kinobetrieb', *Der Kinematograph*, no. 655, 1919.

Author Unknown, 'Die Kinomusik', *LichtBildBühne*, no. 2, 1919.

Author Unknown, 'Engangementspolitik des Kinobesitzers', *Das Kino-Orchester*, no. 11, 1926.

Author Unknown, 'Geräuschmusik auch für kleine Kinoorchester ermöglicht', *Das Kino-Orchester*, no. 20, 1928.

Author Unknown, 'Künstlerdienst mit erschwerenden Umständen', *Der Kinematograph*, no. 579, 1918.

Kleibömer, G., 'Aus dem Reiche der Töne. Kinematograph und Grammophon im Dienste der Kunst', *Der Kinematograph*, no. 127, 1909.

Schmidl (first name unknown), 'Zensurkarten- Jammer', *Das Kino-Orchester*, no. 11, 1925.

Books and articles

Beiche, Michael, 'Musik und Film im Deutschen Musikjournalismus der 1920er Jahre', *Archiv für Musikwissenschaft*, no. 63, H.2. (Stuttgart: Franz Steiner Verlag, 2006), 94–119.

Blum, Carl Robert, *Das Musik-Chronometer und seine Bedeutung für Film-, Theater- und Allgemeine Musikkultur* (Leipzig: F. E. C. Leuckar, 1926).

Dettke, Karl Heinz, *Kinoorgeln und Kinomusik in Deutschland* (Stuttgart and Weimar: J. B. Metzler Verlag, 1995).

Fabich, Rainer, *Musik für den Stummfilm: Analysierende Beschreibung originaler Stummfilmkompositionen* (Frankfurt am Main: Peter Lang, 1993).

London, Kurt, *Film Music: A Summary of the Characteristic Features of its History, Aesthetics, Techniques; and Its Possible Developments* (New York: Faber and Faber, 1936).

Ottenheym, Konrad, *Film und Musik bis zur Einführung des Tonfilms: Beiträge zu einer Geschichte der Filmmusik* (Dissertation. Berlin: Friedrich-Wilhelm-University, 1944).

Robinson, David, *Das Cabinet des Dr. Caligari* (Basingstoke: Palgrave Macmillan, 2013).

Seidler, Walther, *Stummfilmmusik gestern und heute: Beiträge und Interviews anlässlich eines Symposiums im Kino Arsenal am 9. Juni 1979 in Berlin* (Berlin: Verlag Volker Spiess, 1979).

4

Gottfried Huppertz's *Metropolis*: The Acme of 'Cinema Music'

Emilio Audissino

Huppertz is indeed a composer of very high level. [...]
It is really sad that his life was so short
as to prevent him from having a more
extended artistic career, which he would have deserved.
Frank Strobel[1]

There is a basic difference between the music of the silent film era and that of the sound film era. Music for sound films is a material part of the film: the music is printed on the film-strip. Music for silent films was not: the film-strip had no soundtrack. Played live and liable to change from one screening to another, music was rather a part of the film-viewing experience. In Peter Larsen's words, 'the music of the silent film is not film music in the modern sense it is *cinema music*, an external addition to the moving pictures, part of the total *performance* more than part of the film and its narrative.'[2] I take Larsen's distinction here primarily as a periodization concerning the physical placement of the music: in the case of sound films, the music is on the film(strip) and thus called 'film music'; in the case of silent films, the music is in the cinema(s), played live, and thus called 'cinema music'. We may also think of the distinction as one based on the viewers' experience of the music as a part of the film narrative or not – with 'film music' being experienced as more narratively integrated with the film, and 'cinema music' being experienced more as a silence-filler of the film projection rather than a constituent of the filmic system. The combination of the material/period-based and the narrative/function-based interpretation makes the classification a bit more complicated. Indeed, there are cases of music written in the 'cinema-music' period that nevertheless had the narrative importance and integration of the music that would be written in the 'film-music' period – think of Hans Erdmann's original score for Murnau's *Nosferatu* (1921).[3] Gottfried Huppertz's score for Fritz Lang's 1927[4] epic *Metropolis* is the pre-eminent example. Though 'cinema music' according to the material/period-based interpretation, the *Metropolis* score was designed to have such a narrative agency and adherence to the film as to anticipate 'film music' – as if the music were to be closely interlocked to the image track on the film-strip. Its lush orchestration, timbre nuances and leitmotific variety require the original full-orchestra instrumentation to be fully appreciable and meaningful, with much to be lost in a cheaper salon-orchestra or piano reduction, and this requirement is more in line with 'film music' – in the 'cinema-music' era few venues could afford a full symphony orchestra for each show, while in the 'film-music' era the orchestra is recorded only once, dubbed onto the film-strip and ready to be played in any venues (sound equipment permitting). Moreover, with its subtle manipulation of the musical material for narrative effects, Huppertz's score had an unprecedented formal importance in the film. Synch points abound in the score (more than one thousand, as we shall see) and, again, this is something more in line with 'film music' – which is recorded in the controlled setting of a studio take after take until the good one is dubbed onto the film-strip – rather than the more unpredictable live setting of 'cinema music' – which has to be played repeatedly before a live audience.

After a brief account of the vicissitudes that *Metropolis* suffered in its almost ninety years of life, I shall focus on the score's formal contribution to the film, tackling it as if I were discussing a 'film-music' score. In the final section, I shall bring the score back to its 'cinema-music' historical context and shall survey the film's recent screenings with live musical accompaniment.

Reconstructing Fritz Lang's *Metropolis*

It has been estimated that 80 per cent of silent films is lost, perhaps forever.[5] Up to 2008, Fritz Lang's *Metropolis* – that is, the film as Fritz Lang had originally conceived it – was one item in this list.[6] *Metropolis* was planned and advertised as the top *blockbuster* of its time, German cinema's challenge to Hollywood. Shot in seventeen months (310 days and 60 nights of shooting), it featured 36,000 extras, astonishing visual compositions by special-effect wizard Eugen Schüfftan (father of the 'Schüfftan process'[7]), daring camerawork by Karl Freund and Günther Rittau, and futuristic art design by Germany's top architects and designers – Otto Hunte, Erich Kettelhut, and Karl Vollbrecht.[8] Allegedly, it cost a staggering five million Reichsmarks (about €180,000,000).[9] The film opened in Berlin on 10 January 1927 and was 4,189 metres long (about 153 minutes).[10] Immediately after the Berlin screening, Parufamet (the distribution company) deemed the film too long and unsuitable for the American market. From that point onwards, the film underwent a number of edits that drastically altered its form.[11] After the war, the widely circulating version was the MOMA copy – bought in 1936 – which was only 2,530 metres long (91 minutes), three-fifths of the film's original length:

> The forced cuts and the elisions, caused by men, by time, and by chance, resulted in the elimination of not only fundamental parts of the narrative, but also segments that were absolutely indispensable for the film's visual structure and formal design. A film of an exceptional symbolic and figurative richness was thus reduced to a simplistic plot resembling one from a maudlin romance novel.[12]

This *Metropolis* received further visibility when Giorgio Moroder released, in 1984, a new colourized version with a compilation score including songs by Freddie Mercury, Bonnie Tyler and others.

Gottfried Huppertz's original score suffered a similar fate. The composer prepared a new version of the score to conform to the first 3,241 metre re-edit. There is no evidence that Huppertz was asked to rework the music for the second re-edit (the Paramount version prepared for the American market). This would have entailed new costs and would have been quite pointless anyway, as the American distribution typically would replace the European scores with newly commissioned ones – a process called 'localization'.[13] Indeed, a reviewer, who had seen the original version, thus wrote about the Paramount version:

> A few motifs remain from the music by Gottfried Huppertz, and he is still credited as the composer, but with these exceptions, the accompaniment has switched to the tried and tested medley of Chopin, *Freischutz* and *Traviata* along with popular songs from the Kinothek. [...] The difference between the two versions is comparable to the difference between the Tannenberg monument and the war memorial stone in Alt-Ruppin.[14]

Like the original film, after the January 1927 première Huppertz's score soon sank into oblivion. Indeed, the score must have been completely forgotten by 1984 (the year of the release of Moroder's version), Thomas

Elsaesser provided an utterly distorted and yet confident opinion of what the original music sounded like:

> *Metropolis*, too, originally had a score, composed by Gottfried Huppertz, which in the 1970s was used to showcase the restored archive version on German television. But Huppertz's music – a medley of Liszt, Wagner, and Tchaikovsky performed mainly on the piano – is salon music, curiously irrelevant to the film because it is devoid of associations other than to the pomposity of first-night audiences at mid-1920s movie palace premieres.[15]

Fritz Lang's *Metropolis*, the original edition, soon became the Holy Grail of film restorers, and starting in the late 1960s, Enno Patalas – film historian at the Münchner Stadtmuseum Filmmuseum – launched an international archival research to reconstruct it.[16] The 1987 Patalas edition brought the film to 3,170 metres (115 minutes), filling the still missing parts with production stills and summarizing intertitles. In 1998, Martin Koerber – at the Deutsche Kinemathek – followed in Patalas's footsteps and prepared a new version. Presented in 2001, the Koerber edition, with its 3,590 metres (130 minutes), was announced as the final reconstruction. With the still-missing 600 metres deemed lost forever, in 2008 the film historian Fernando Peña surprisingly retrieved a copy of the unabridged 1927 version at the Museo del Cine 'Pablo C. Ducros Hicken' in Buenos Aires.[17] A new reconstruction could begin, which would produce the 2010 Argentinian edition (147 minutes), the closest to the original that we have.

Among the materials that Patalas had found, there were the original scenario, the censorship cards (*zensurkarte*), and Huppertz's musical sketches that came with a number of detailed annotations referring to the film's scenes and intertitles. The 1987 Patalas edition used Huppertz's materials as reference points for the reconstruction, but had no musical accompaniment. When the Koerber 2001 edition was released on DVD, the original score was included as a full-orchestra recording arranged to fit this shorter version.[18] Yet, the music had a really important final word in the 2010 Argentinian edition. By comparing the 2001 reconstruction with the 1927 version finally retrieved in Buenos Aires, Koerber and his associates noticed that there were differences in terms of duration and order of the shots. They realized that, to reconstruct the film as faithfully as possible, Huppertz's score was the ideal – and possibly only – fine-tuning instrument.[19] As a consequence, Koerber and Anke Wilkening hired conductor Frank Strobel not only to reconstruct the full score, but also as the third specialist in the editorial team that would produce the 2010 Argentinian edition: 'That score is the only complete document from the 1927 première [...]. The music played a big role right from the beginning because the film's editing was based to the score itself',[20] explains Strobel.

The full orchestral score was reassembled from Huppertz's short-score sketches, his handwritten orchestrations, a salon orchestra and a piano arrangement published at the time.[21] Since the first sixty-one pages of Huppertz's 701-page full-orchestra manuscript were missing, a re-orchestration from the other existing sources was in order.[22] A digitized copy of the film was run on a computer and checked against a MIDI version of the score played at the right tempo. If the visual events hit the musical events as indicated in Huppertz's annotations, then it meant that the film had been reconstructed in the correct manner.

Gottfried Huppertz's *Metropolis* score

Gottfried Huppertz (11 March 1887, Cologne, Germany – 7 February 1937, Berlin), formerly an actor and an opera singer, started his career as a film composer with *Die Nibelungen* (1924), Fritz Lang's two-part epic.[23] Huppertz's involvement in *Metropolis* started in the early stages of the film's production, with the aim of penning a score consisting of all-original music, not compiled from anthologies.[24] He would be on the set to develop the themes for each character as they took form in front of the cameras, and also played parts of the music for the actors to bring them into the proper mood.[25] As a result, Huppertz's score is so integral to the film as to be a very rare if not unique case in the silent cinema repertoire.[26] Before moving on to the music itself, a quick summary of the film's story might prove useful.

Metropolis is set in an unspecified point in the future, in a hyper-technological city ruled by the technocrat Joh Fredersen. The powerful machines that keep the city running are operated by a huge mass of workers, who fight the bleakness of their exploited lives by secretly gathering in regular meetings in the city's underground, where Maria, a young girl, preaches of the coming of a Mediator who soon will bring a time of changes. Freder, Fredersen's spoilt son, happens to meet Maria and falls in love. While looking for her in the workers' quarters, he discovers their unbearable life conditions. Shocked, he resolves to get things changed and talks to his father. When Freder meets Maria again, she realizes that he is the long-awaited Mediator. In the meantime, Fredersen decides to halt his son's involvement with Maria and the workers. He visits Rotwang the scientist and there he discovers a memorial to Hel, Fredersen's late wife who had previously been Rotwang's beloved. Rotwang is obsessed with the woman and has been building a robot in order to create a simulacrum of her. Fredersen asks Rotwang to give Maria's semblance to the robot. Maria is abducted and replaced by the False Maria, who begins to discredit the real one by acting outrageously. Yet, Rotwang seizes the occasion to take his revenge on Fredersen. He instructs the False Maria to lead the workers into a violent insurrection, with the aim of destroying Fredersen's city. The workers devastate the machines, unintentionally causing a flood in their quarter. Believing their children have died in the flood as a result of False Maria's incitements, the workers decide to burn her on the stake. In the meantime, Maria is freed and, with Freder's help, rescues the children. The False Maria is revealed to be a robot and Rotwang is identified as the culprit. After a chase on the cathedral's roof, Rotwang falls to his death. Fredersen arrives at the site and finally realizes the terrible consequences of his wrongdoing. In front of the cathedral's portal, Grot, the chief worker, and Fredersen are encouraged by Freder to reconcile and finally shake hands. 'The mediator between head and hands must be the heart!', thus reads the inscription that closes the film.

As to the score's stylistic traits,[27] the main compositional technique is the leitmotif, i.e. a recognizable musical idea clearly associated with some character or concept in the film. The leitmotif here is already extensively employed in its shortened and more functional form as it would be in the Golden Age of Hollywood music (1933–58) – a 'condensed leitmotif' as Royal S. Brown calls it.[28] In Peter Larsen's words:

> [T]he music can be described as a montage of leitmotifs. In stage music based on leitmotifs there will normally be a musical tension between 'figure' and 'ground'. The leitmotifs are integral parts of the overall movement, but at the same time can be distinguished as independent elements, and when they are experienced in this way as 'figures', the rest of the music functions as a kind of more neutral, unmarked 'background'. The characteristic thing about Huppertz's composition, however, is that the tension between figure and ground is neutralized or, rather, there is no ground, there are only figures. The music consists almost entirely of leitmotifs.[29]

This definition can apply to the classical Hollywood music as well, in Max Steiner's use of the leitmotif technique, for example. Indeed, Huppertz's score boasts a number of leitmotifs that is no second to Erich

Wolfgang Korngold's Hollywood scores. Another striking 'Hollywood-like' stylistic trait of the score is the very tight music/visuals correspondence. There are instances of what would be called 'Mickey-Mousing' technique: visual actions are mirrored by music actions in tight synchronism. For example, harshly dissonant chords give voice to the steaming sirens signalling the shift in the machine room; when Freder lifts his head and notices the arrival of Maria in the Eternal Gardens, a violins tremolo highlights this action; clashes of cymbals punctuate each slash of Death's scythe at the end of the False Maria dance at Yoshiwara; the whole sequence of Rotwang chasing Maria in the catacombs is a piece of tightly synchronized action music, with the grim sight of some skeletons appropriately marked by xylophone touches – a reference to Camille Saint-Saëns's *Dance Macabre* (1874). The score's structure is so close to the film's structure that 1,028 synchronization points are indicated in Huppertz's manuscripts.[30]

The score's general language is rooted in the late-romantic nineteenth/early twentieth-century music, with echoes of Wagner in the harmonic treatment (for example, in Freder's theme), Mahler in the orchestral voicing (for example, in the film's finale), and Richard Strauss in the melodic lines – especially in the brass raising phrases, as in the opening fanfare. Yet, the score is also tinted with contemporary dialects: for example the Yoshiwara foxtrot – using saxophones and stemming from the 1920s dance and variety music, sounding like a Kurt Weill cabaret song – and the 'Machine Theme', which displays the traits (pounding ostinatos, clashing dissonances, onomatopoeic writing) of the 1920s Futurism-inspired 'Machine Music' – examples include Arthur Honegger's *Pacific 231* (1923), Sergei Prokofiev's *Symphony No. 2 in D Minor* (1925) and George Antheil's *Ballet Mécanique* (1926). The score was conceived for full symphony orchestra – double woodwinds with saxophones, ten brasses, timpani and percussions, piano, organ, celesta, harp and strings – and at the 1927 première was performed by a 66-piece ensemble.[31]

The score performs a set of functions that go well beyond the standard of the music composed for the silent films. Music is not merely illustrative (setting the locale and mood of a scene) or a rhythmic boost to the action (fast-paced battle music to add dynamism to a battle scene), which were the central concerns of most handbooks and anthologies for silent film accompaniment. There are indeed proficient cases of musical illustration – during the human sacrifice to Moloch, loud timpani beats and parallel-moving fifths in the brass provide archaic sonorities for this pre-Christian rite[32] – cases of mood music – the music accompanying the scenes in Rotwang's lab is appropriately sinister, with muted trumpets sforzandos and pizzicato low strings – and cases of effective action music – for example, the frantic piece accompanying the rescue of the workers' children. There are occurrences of dance music used to set the ambience: a catchy foxtrot for the Yoshiwara sequences and an elegantly vacuous waltz for the Eternal Gardens, the recreational retreat of the oligarchs' sons. Yet, Huppertz's score is also subtler. For example, when Josaphat loses his job as Fredersen's secretary, we hear the 'Workers' Theme' as we see his anguished face. The music here communicates his thoughts: being fired by Fredersen means no other job opportunities than low-level employment in the life-consuming machine room.

The music is highly connotative, providing hints to the film's implicit meanings and disclosing a number of conceptual connections. Besides its apparently simplistic love and class-struggle story, the film has a rich network of visual and thematic symbolism. Expressionism – with the typical Doppelgänger motif – meets New Objectivity – with the depiction of the modern city life. The Modern blends with the Ancient – the Gothic architecture of the cathedral is embedded within the city's futuristic skyline; Rotwang's work is a mix of science and alchemy; modern advanced technology relies on the most ancient form of workforce: slavery. Visual symbols are scattered throughout the scenery – alchemical spirals and occultist pentacles, but also Christian crucifixes – and even the characters' names themselves are highly symbolical – Freder is the son of Joh Fredersen, but his father's surname sounds like 'Fredersohn', 'Son of Freder', as to indicate that the powerful lord of Metropolis is to be given a new birth by his own son Freder; Freder the Mediator sounds like 'Frieden': 'Peace'.[33] Both Freder and Maria incarnate Christian figures: Freder is the Mediator (the Saviour) and Maria (Mary) is she who brings him into the (workers') world, interceding from a meeting venue in the catacombs (resembling those of the persecuted early Christians). Music is a key component of this symbolical network. For examples, through a clever manipulation of the themes, music stresses Freder's maturation from a spoilt rich kid to a socially aware activist. We hear Freder's theme for the first time in a bragging horns rendition during the scene at the Stadium: Freder here is killing some time in a running competition with his peers. Then he meets Maria and things begin to change in his soul. When Josaphat, his father's secretary, is sacked and then attempts suicide, Freder stops the man's hand and offers him his help. Here, Freder's Theme reappears but in a significantly different version: played with a warmer and sincerely sympathetic élan, backed by Puccini-like high-register violins giving the brass a heartfelt glow, a soul. Another instance is to be found after the accident in the machine rooms, where a number of workers die: the 'Dies Irae' is heard here for the first time. Right after this scene, we cut to a majestic view of the skyscrapers and busy highways of Metropolis, the music reprising the opening fanfare but adding a secondary ascending phrase in a triumphant major mode. It is a celebration of technology and progress. Then, we cut to Fredersen's office, the technocrat who tyrannically rules over the place. 'Fredersen's Theme' is introduced here. It is a menacing minor-mode and descending version of the secondary phrase that we have just heard over the city views. The musical sequence reinforces the conceptual connection of the three scenes: the other side of the coin of the technological triumph of Metropolis (ascending major-mode fanfare) is a tyrannical regime (descending minor-mode fanfare), and this technology is the fruit of the rough exploitation of human beings, which sooner or later is bound to lead to catastrophic consequences ('Dies Irae').

Musical quotations, in particular, are used to suggest implicit meaning. The 'Dies Irae' – the medieval sequence from the Mass for the Dead, amply used in the history of music to imply impending danger – is recurrently featured, notably in the scene where the monk in the cathedral announces that the Armageddon is nigh. Interestingly, the monk is played by Fritz Rasp, the same actor who plays Der Schmale, the private detective sent by Fredersen to spy on his son. Why this strange choice of having Rasp play these two different and unrelated characters? The first time Der Schmale is introduced, we hear the 'Dies Irae', and so do we each time he appears. The 'Dies Irae' comes with his traditional gloomy connotations, and this association with death is confirmed in its first appearance during the accident in the machine room. The accident prefigures the disaster (the 'Day of Wrath') at the end of the film, when the city is nearly destroyed. How have we arrived at this disaster? The main character responsible is Fredersen, who asked Rotwang to create the False Maria to impede his son's progressive actions. And Fredersen's first step in his plot against his son is to send Der Schmale on his trail. Hence, the spying of Der Schmale (musically signalled by the 'Dies Irae') is actually the ignition of a chain reaction to the 'Day of Wrath'. The music offers an explanation to this otherwise strange double-character casting choice. Another musical quotation is 'La Marseillaise', which is repeatedly featured when the False Maria leads the mob to the destruction of the machines. The music here provides a negative comment of bottom-up revolutions in general by presenting the famous French anthem in a rhythmically distorted minor-mode rendition. Finally, there is an isolated melody that appears twice at the end of the flood sequence, when Maria (Mary), with the help of Josaphat (Joseph) and Freder (the Messiah), manages to save the children leading them upstairs. The melodic contour of this theme is closely reminiscent of the traditional Catholic Marian chant 'Mira il tuo popolo, o bella signora'. Identifying this melody as a quotation aimed at confirming Maria as a symbol of the Virgin Mary does not seem too far-fetched an interpretation, given Lang's Catholic upbringing and the vast popularity of this chant.[34]

For reasons of space, we cannot look further into the score's complexity and its innovative nature. However, these few insights should be enough to show how important it is that not only Lang's film has been reconstructed, but Huppertz's original score too. Despite being a 'silent' film, *Metropolis* is a thoroughly designed audio-visual artwork and the film cannot be appreciated – or fully interpreted – without its complementary score.

Huppertz's *Metropolis* today

The world première of the reconstructed film (and score) – the Argentinian 2010 version – took place in Berlin on 12 February 2010 at the Friedrichstadt-Palast, with Strobel conducting the Rundfunk-Sinfonieorchester Berlin, with a simultaneous outdoor projection on a special screen on the Brandenburg Gate.[35] The live accompaniment of a film requires a conductor able to elicit a good musical performance from the orchestra and, at the same time, to keep the live music in synch with the film. Strobel conducted free-style, without technical aids:

> Similar to an operatic performance, I need to be able to shape and structure a work and to uphold its tension, without having the feeling of being at the movie's mercy. And it is especially important to me that a performance take place without any additional technical aids, such as a click track in my ear or a visible time code on the monitor – for musicality's sake. Synchronicity can be achieved by a precise knowledge of the film, plentiful (mostly more than a thousand) synchronicity pointers in the score, exact tempo specifications (preferably metronomic indications) and the aforementioned feeling for movement and mounting in the picture.[36]

In the last decade there has been a blossoming of 'multi-media film' shows, with increasingly more orchestras accompanying live not only silent but also sound films – starting with John Williams conducting live his score to *E.T. The Extraterrestrial* (Steven Spielberg, 1982) in 2002, with recent examples including *The Wizard of Oz* (Victor Fleming, 1939), *West Side Story* (Jerome Robbins, Robert Wise, 1961), the *Lord of the Rings Trilogy* (Peter Jackson, 2001–3), and *Titanic* (James Cameron, 1997).[37]

> Silent cinema, for many years, used to be a repertoire for few specialists and fans. Today, in 2013, we can say the contrary: the silent film as a show including a live performance of its original score or – as is often the case – of a newly commissioned score has firmly established itself as an artistic expression akin to opera, musical theatre, or ballet.[38]

Indeed, since the 2010 Berlin première, *Metropolis* as a 'multi-media film'[39] – a package of projected moving images and live music – has travelled throughout the world, even landing at La Scala, Italy's 'sacred temple' of opera, and is still being presented regularly.[40] Huppertz's music has been unanimously praised. *The Wall Street Journal* wrote: 'For those familiar only with Moroder's music, featuring Freddie Mercury and Pat Benatar, Huppertz's operatic score is a revelation',[41] whilst *The Independent* added that the previous Moroder version 'would feel downright archaic compared with the vitality of the *Metropolis* released last week, which is the closest we've yet seen to Lang's original vision. [...] More than an average restoration, *Metropolis* is one of the best new films of the year – and all the better without Billy Idol.'[42] Besides the theatrical screenings, the 2010 edition of *Metropolis* has also circulated as a DVD, and a 75-minute selection from the score was released on CD.[43] *Metropolis* also holds the record for being the first silent film to be released on Blu-ray.[44]

Besides the general consensus about the score's quality, a curious retroactive sort of psychological projection has occurred. Many observers have noted how Huppertz's score sounds like Hollywood music: 'Inhabiting as it does the musical world of Korngold, the score should appeal to all lovers of the high romantic in music',[45] or, 'The score, by the undeservedly unknown Gottfried Huppertz [...] reflects a similar blend of 19th century and modernist archetypes and anticipates the take-up of Wagner's *leitmotif* technique by generations of film composers to shadow and reinforce filmic narrative',[46] or 'Gottfried Huppertz's 1920s orchestral score [sounds] uncannily like John Williams.'[47] Of course, it is Hollywood music that sounds like Huppertz's, since Huppertz predates Hollywood music.

Given the similarity, can it be argued that the Huppertz *Metropolis* score may have had some sort of influence on the classical Hollywood music? Quite certainly not. The film distributed in the United States was the Paramount abridged version and, as we have seen, it was probably accompanied by a compilation score, maybe including only one or two rearranged Huppertz themes. Also, the 1927 shellac discs with a suite from the score had a narrating voice in German – by Lang himself – and I have found no evidence of an American release. Similarly, it is highly improbable that some of the founding fathers of Hollywood music might have watched the original version in January 1927 and got in touch with its music. Max Steiner was already a US resident, working on Broadway musicals; the American-born Alfred Newman was in his homeland; Erich Wolfgang Korngold was busy in Vienna, working on his forthcoming opera *Das Wunder der Heliane* (*The Miracle of Heliane*) – premièred on 7 October 1927 – and had not yet turned his attention to films; German-born Franz Waxman might have been in town at the time, but there is no evidence that he might have seen the film, and he never mentioned Huppertz's score. Huppertz sounded like Hollywood because his training took place in the same cultural milieu of Korngold, and their idioms are therefore similar. It is likely that the resurfaced score has made such an impression on present-day audiences exactly because, compared to other original scores from the silent era, it sounds and functions much more like the Hollywood music we have become accustomed to. What can be said speculatively is that had Huppertz migrated to Hollywood as Lang, Erich Pommer, Franz Waxman and other illustrious countrymen did, he might have been one of the founders of Hollywood music, as he did possess the ideal musical style and sense of drama. What can be said with confidence is that his score to *Metropolis* – 'cinema music' designed to be so tightly interlocked with the film as to prefigure 'film music' – is quite enough an achievement to secure him a place in the history of film composers.

I would like to close with a somewhat provocative remark that, hopefully, might be useful to prompt further reflections on silent films and their original music. There has been a silent cinema revival in the last three decades, including screenings with live musical accompaniments. In most cases the films are accompanied with newly composed (or even improvised) solutions that can span from symphonic to noise music. The main reason for using newly composed scores is that original scores are often missing. Another reason is to update the musical style in order to be more in line with the taste of contemporary audiences.

Metropolis too was musically updated, most notably in the 1980s with Moroder's pop compilation, and in the 1990s with the Alloy Orchestra's electro/industrial revisitation, two successful versions that catered to audiences that might have not enjoyed a more traditional symphonic accompaniment. If no original score is available for a silent film – or if the original score was intended merely as a silence-filler not really connected with the film's narrative – then any musical accompaniment is legitimate. Yet, now that we have the original Huppertz score, such operations become quite questionable. As shown in the previous pages, Huppertz's score was conceived as an integral part of the film, perhaps more than any other silent film score – so integral that the score was the key aid in the reconstruction of the film itself. To replace it with another form of musical accompaniment equals to arbitrarily devoid the film of very important narrative hints and connotations that are carried or revealed by the original score. Neither Moroder's nor the Alloy Orchestra's music respond to the film as deeply and extensively as Huppertz's music does. To replace Huppertz's score is like replacing Prokofiev's *Alexander Nevsky*, Morricone's *Once Upon a Time in the West*, or Williams's *Close Encounters of the Third Kind*, three scores that were the result of a close collaboration between the director and the composer and whose replacement would probably be felt as unrespectful of the integrity of the author's vision, perhaps even outrageous and preposterous. The same should be felt with *Metropolis*: Huppertz's score was the result of a similarly close collaboration with Lang and it was so precisely designed to be intertwined with the visuals and the narrative that it should be treated as if it were interlocked 'film music' instead of replaceable 'cinema music'.[48] Huppertz's score is *the* music for *Metropolis*.

Notes

1. Paolo Eustachi, 'Frank Strobel e *Metropolis*', *Colonne Sonore*, 4 August 2013, online, accessed 2 February 2015, http://www.colonnesonore.net/contenuti-speciali/interviste/2695-frank-strobel-a-metropolis.html, translated from Italian.
2. Peter Larsen, *Film Music*, trans. John Irons (London: Reaktion Books, 2007), 26.
3. Erdmann's reconstructed score can be heard in the *Nosferatu* restoration by the 'Friedrich Wilhelm Murnau Stiftung' released as a DVD by Eureka, 'Masters of Cinema Series', #64, 2007.
4. I prefer to date *Metropolis* to 1927 because its première took place in January of that year. Many date the film to 1926, because it was approved by the censorship board in November 1926.

5. Paolo Cherchi Usai, *Silent Cinema: An Introduction* (London: BFI, 2000, 2nd edn.), 10.
6. A reference book in English on the *Metropolis* case – covering the production, the distribution, the restoration (up to the year 2000), and the interpretation of the film – is *Fritz Lang's Metropolis: Cinematic Visions of Technology and Fear*, ed. Michael Minden and Holger Bachmann (Rochester, NY: Camden House, 2000).
7. See Katharina Loew, '14 September 1922: Schüfftan Process Reconciles Artistic Craftsmanship with Demands of Entertainment Industry', in *A New History of German Cinema*, ed. Jennifer M. Kapczinski and Michael D. Richardson (Rochester, NY: Camden House, 2012), 111–16.
8. See Enno Patalas, *Der Fall Metropolis*, documentary film, Germany, Friedrich-Wilhelm-Murnau-Stiftung and Transit Film, 2003.
9. The figure was divulged by the UFA press office – 'probably inflated for promotional purposes', Paolo Bertetto, *Metropolis* (Turin: Lindau, 2001, 3rd edn.), 11. Another hypothesis places the figure around 3,500,000 Reichsmarks (Artem Demenok, *Die Reise Nach Metropolis*, documentary film, Germany, Südwestrundfunk, 2010). In any case, it was probably the most expensive European film at the time.
10. In silent cinema, the frame rate at which the film was shot was not standardized, and it could be different from the rate at which the film would be projected – see Kevin Brownlow, 'Silent Films – What Was the Right Speed?', in *Early Cinema: Space, Frame, Narrative*, ed. Thomas Elsaesser and Adam Barker (London: BFI, 1990), 282–90. Since often there is no record of the original rate, it is now a matter of guesswork. At a sound-film projection rate of 24 f/s (the speed eventually chosen for the 2010 reconstruction), the 4,189-metre *Metropolis* has a running time of 153 minutes. This seems a good approximation, as in Gottfried Huppertz's manuscripts the music is apparently written with a 26 f/s rate in mind – see Frank Strobel, 'Rekonstruktion und Originalmusik von *Metropolis*', FrankStrobel.de, accessed 21 January 2015, http://www.frankstrobel.de/texte-von-frank-strobel/metropolis.html. Two contemporary reviews – mentioning a film's running time of circa two and a half hours – also confirms this as an appropriate projection rate: see Holger Bachmann, 'The Production and Contemporary Reception of *Metropolis*', in *Fritz Lang's Metropolis*, ed. Minden and Bachmann, 26 n66.
11. On the film's troubled history of acts of tampering and attempts of recovery, see Bertetto, *Metropolis*, 11–45; Thomas Elsaesser, *Metropolis* (London: BFI 2000), 30–42; Minden and Bachmann (eds.), *Fritz Lang's Metropolis*, 109–57.
12. Bertetto, *Metropolis*, 12–14, translated from Italian.
13. See Sergio Miceli, *Musica per film: Storia, Estetica, Analisi, Tipologie* (Lucca-Milan: LIM/Ricordi, 2009), 114–33.
14. Quoted in Enno Patalas, 'The City of the Future – A Film of Ruins: On the Work of the Munich Film Museum', in *Fritz Lang's Metropolis*, ed. Minden and Bachmann, 115.
15. Thomas Elsaesser, 'Innocence Restored? Reading and Rereading a "Classic"', in *Fritz Lang's Metropolis*, ed. Minden and Bachmann, 125. Elsaesser probably referred to some TV broadcast of *Metropolis* in which a typical silent-film

compilation score was passed off as Huppertz's music, possibly the same compilation score mentioned by the 1927 reviewer I have previously quoted. In his 2000 monograph on the film, Elsaesser never mentions Huppertz but only Moroder: it can then be argued that the Huppertz score really started to be rediscovered only with the release of the 2001 Koerber edition.

16. An account and shot-by-shot description is in Enno Patalas, *Metropolis in-aus Trümmern* (Berlin: Dieter Bertz, 2001).
17. On the Buenos Aires discovery, see Karen Naundorf, 'The Metropolis Mystery', *Sight and Sound*, vol. 18, no. 9 (2008): 26–9, and *Die Reise Nach Metropolis*.
18. '"The music was also the basis for the 2001 reconstruction." But back then, Mr. Koerber complained, he needed to bend and break the music to fit the film.' A. J. Goldmann, '*Metropolis* Now', *The Wall Street Journal*, 19 February 2010, online, accessed 22 January 2015, http://www.wsj.com/articles/SB10001424052748704804204575069383586138. The music was re-orchestrated, arranged and conducted by Berndt Heller with the Rundfunksinfonieorchester Saarbrücken.
19. A fundamental intermediate step was the 2005 study version prepared by Patalas for the Universität der Künste, Berlin (released on DVD by the university's 'Filminstitut'). Before the Argentinian materials had been found, Patalas assembled the available parts alongside the complete score (in a two-piano recording), leaving black visual gaps as an indication of the sections whose material was missing.
20. Goldmann, '*Metropolis* Now'.
21. Strobel, 'Rekonstruktion und Originalmusik von *Metropolis*', online.
22. Ibid.
23. For Huppertz's life and career see Aitam Bar-Sagi, 'Gottfried Huppertz', Film Music Museum, online, accessed 20 January 2015, http://fimumu.com/huppertz.
24. On anthologies and cue sheets, see Rick Altman, *Silent Film Sound* (New York: Columbia University Press, 2004), 249–70, 345–66.
25. Hans-Jürgen Linke, 'Interview mit Dirigent Frank Strobel: Von der Macht der Musik im Stummfilm', *Frankfurter Rundschau*, 28 January 2010, online, accessed 22 January 2015, http://www.fr-online.de/musik/interview-mit-dirigent-frank-strobel-von-der-macht-der-musik-im-stummfilm,1473348,2824380.html. While this practice of having music played on-set to inspire the actors' performance was not unusual at all in the silent era, what was highly unusual was having the composer of the original music playing parts of the work-in-progress score.
26. There are other instances of close composer/director collaborations in the silent era, for example those between Joseph Carl Breil and David W. Griffith for *The Birth of a Nation* (1915) or Abel Gance and Arthur Honegger for *Napoleon* (1927). Yet, the former score was only partially original, still employing much repertoire music as was customary in the 1910s, while the latter is lost in its full form and the surviving parts do not seem to suggest that the music was designed to closely adhere to the film narrative and visual actions.
27. 'Style' is not used here in its narrow sense, i.e. the 'language' of the music – baroque style, classical style, romantic style, etc. For film music, I adopt a

definition that takes into account four stylistic traits: language, techniques, musical means and functions. See Emilio Audissino, *John Williams's Film Music: Jaws, Star Wars, Raiders of the Lost Ark, and the Return of the Classical Hollywood Music Style* (Madison, WI: University of Wisconsin Press, 2014), xxiv–xxvi.

28. Royal S. Brown, *Overtones and Undertones: Reading Film Music* (Berkeley, CA: University of California Press, 1994), 97–118.
29. Larsen, *Film Music*, 62.
30. Strobel, 'Rekonstruktion und Originalmusik von *Metropolis*', online.
31. Larsen, *Film Music*, 48. To accommodate the needs of those screening venues that could not afford a full symphony orchestra, Huppertz also prepared one version for piano solo and one reduction for salon orchestra.
32. The scene can be quite safely said to be inspired by the famous scene of the Phoenician human sacrifice to Moloch in *Cabiria* (Giovanni Patrone, 1914).
33. Readings of the allegories and connotations in *Metropolis* can be found in Tom Gunning, *The Films of Fritz Lang: Allegories of Vision and Modernity* (London: BFI, 2000), 52–83; Bertetto, *Metropolis*, 73–209; Elsaesser, *Metropolis*, 42–56; and Minden and Bachmann (eds.), *Fritz Lang's Metropolis*, 159–285.
34. Illustration – 'Mira il tuo popolo':

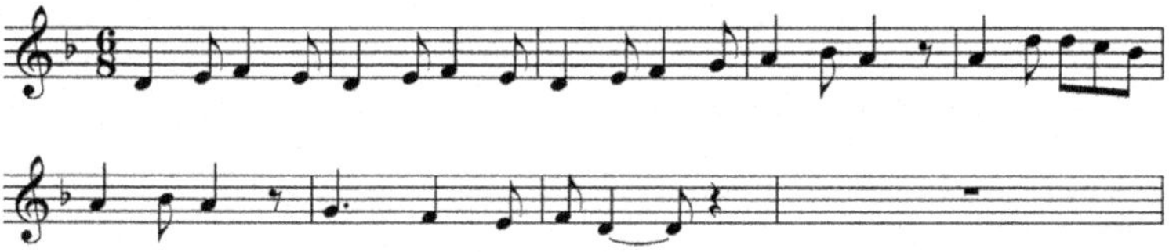

The attribution is uncertain. 'Mira il tuo popolo' could have been penned by Saint Alfonso Maria De' Liguori in the eighteenth century, although it is also ascribed to an anonymous author from Pisa – see St Alfonso Maria De' Liguori, *Opere ascetiche. Vol. 1: Introduzione generale*, ed. O. Gregorio, G. Cacciatore and D. Cappone (Rome: Edizioni di Storia e Letteratura, 1960), 16. The popularity of the chant has spread well beyond Italian borders and it is highly probable that it was known among the Catholics in Germany at the time. Thanks to Father Alberto Casella for his advice.

35. Anon., 'Metropolis', Berlinale, 2010 edition, online, accessed 2 February 2015, https://www.berlinale.de/en/archiv/jahresarchive/2010/02_programm_2010/02_Filmdatenblatt_2010_20105927.php. Strobel's work on *Metropolis* has produced two editions of the reconstructed score. On the one hand, the 'Ur-text' to be used for study purposes, and an 'Aufführungs' (performance) version, to be used to play the music live to the film (Linke, 'Interview mit Dirigent Frank Strobel', online). Both editions are published by the Europäische Film Philharmonie Institute: http://www.filmphilharmonie.de.
36. Tobias Fischer, 'Fifteen Questions to Frank Strobel', Tokafi, online, accessed 2 February 2015, http://www.tokafi.com/15questions/15-questions-to-frank-strobel.
37. On the forms and formats of film-music concerts, including multi-media films, see Emilio Audissino, 'Overruling a Romantic Prejudice. Forms and Formats of Film Music in Concert Programs', in *Film in Concert: Film Scores*

and their Relation to Classical Concert Music, ed. Sebastian Stoppe (Glücksstadt: VWH Verlag, 2014), 25–44.

38. Eustachi, 'Frank Strobel e *Metropolis*', online, translated from Italian.
39. On 'multi-media' and film with live accompaniment, see Emilio Audissino, 'Film Music and Multimedia: An Immersive Experience and a Throwback to the Past', in *Sounds, Music & Soundscapes, Jahrbuch Immersiver Medien* (Yearbook of Immersive Media), ed. Patrick Rupert-Kruse (Marburg: Schüren Verlag, 2014), 46–56.
40. At the time of writing this (February 2015), *Metropolis* is scheduled to be screened with the live Huppertz score in Germany (March, in Dortmund); in Northern Ireland (March, in Belfast); in England (April, in Leeds); in Italy (September, in Milan). For an up-to-date database of all the forthcoming film-music concerts all around the world – including films with live music – see the 'Movies in Concert' website, accessed 3 February 2015, http://moviesinconcert.nl.
41. Goldmann, '*Metropolis* Now'.
42. Jonathan Romney, 'Metropolis, Fritz Lang, 145 mins (PG)', *The Independent*, online, accessed 3 February 2015, http://www.independent.co.uk/arts-entertainment/films/reviews/metropolis-fritz-lang-145-mins-pg-2076981.html.
43. The DVD/Blu-ray was released by Kino in the USA and by Eureka (Masters of Cinema Series) in the UK in November 2010. The music CD was released in May 2011 by Capriccio (C5066).
44. Linke, 'Interview mit Dirigent Frank Strobel', online.
45. James McCarthy, 'Review of Gottfried Huppertz's *Metropolis*', *Limelight*, 17 August 2011, online, accessed 2 February 2015, http://www.limelightmagazine.com.au/Review/267148,metropolis-original-motion-picture-score-berlin-radio-orchestra/strobel.aspx.
46. Peter McCallum, 'Symphony Soars and Saws in Vivid Saga', *The Sidney Morning Herald*, 31 October 2011, online, accessed 2 February 2015, http://www.smh.com.au/entertainment/music/symphony-soars-and-saws-in-vivid-saga-20111030-1mqe1.html.
47. Romney, '*Metropolis*, Fritz Lang, 145 mins (PG)'.
48. Despite the clearly expressed preference of the F. W. Murnau Stiftung (holder of the rights to the restored 2010 version) to present the film with its original score, the US première took place at the Graumann's Chinese Theater in Los Angeles on 25 April 2010 with a new score by the Alloy Orchestra: 'Alloy's unconventional approach helped make silent films popular again in the 1990s. [...] But some purists prefer to see "Metropolis" with its original score from 1927. In fact, the Turner Classic Film Festival could have chosen to feature that score instead of the Alloy's, according to Donald Kim, head of Kino International, the company that distributes "Metropolis" in America. "People running the show at Turner love the Alloy Orchestra, and that's all they wanted," Kim said. But the people handling the film's painstaking, year-long restoration in Germany wanted the original score for the January première in Berlin. They also insisted it be included with the upcoming DVD of "The Complete Metropolis."' Andrea Shea, 'Retro "Metropolis": Boston Trio Re-Scores A Sci-Fi Classic', *WBUR News*, 23 April 2010, online, accessed 20 April 2015, http://www.wbur.org/2010/04/23/alloy-metropolis. The problem with

original music in silent films has been addressed more assertively by the Roy Export S.A.S./Charlie Chaplin Estate, which forbids any Chaplin post-1918 silent films to be accompanied with any other music than the original: 'We never allow our films to be screened without the full symphonic score that Chaplin himself composed, either on the film's actual soundtrack or played live by an orchestra. Any new arrangements or any piano or organ reductions of the original score are not authorized. We are sure that you will understand our policy that as Chaplin himself composed music for his own films, they should only be screened with that music as he wanted it. It isn't for us to change this. [...] Chaplin did not compose music for the many films he made before 1918. These can happily be accompanied by the music of your choice, so musicians have a free hand to compose the music that they think fits Chaplin's humour.' 'FAQ Live Orchestral Screenings of Charlie Chaplin Films', online, accessed 20 April 2015, http://cineconcert.charliechaplin.com/faq.

Bibliography

Altman, Rick, *Silent Film Sound* (New York: Columbia University Press, 2004).

Anon., 'Metropolis'. https://www.berlinale.de/en/archiv/jahresarchive/2010/02_programm_2010/02_Filmdatenblatt_2010_20105927.php.

Audissino, Emilio, 'Film Music and Multimedia: An Immersive Experience and a Throwback to the Past', in *Sounds, Music & Soundscapes, Jahrbuch Immersiver Medien [Year Book of Immersive Media]*, ed. Patrick Rupert-Kruse (Marburg: Schüren Verlag, 2014).

Audissino, Emilio, *John Williams's Film Music: Jaws, Star Wars, Raiders of the Lost Ark, and the Return of the Classical Hollywood Music Style* (Madison, WI: University of Wisconsin Press, 2014).

Audissino, Emilio, 'Overruling a Romantic Prejudice: Forms and Formats of Film Music in Concert Programs', in *Film in Concert: Film Scores and their Relation to Classical Concert Music*, ed. Sebastian Stoppe (Glücksstadt: VWH Verlag, 2014).

Bachmann, Holger, 'The Production and Contemporary Reception of Metropolis', in *Fritz Lang's Metropolis: Cinematic Visions of Technology and Fear*, ed. Michael Minden and Holger Bachmann (Rochester, NY: Camden House, 2000).

Bar-Sagi, Aitam, *Gottfried Huppertz*. Film Music Museum. http://fimumu.com/huppertz.

Bertetto, Paolo, *Metropolis*, 3rd edn. (Turin: Lindau, 2001).

Brown, Royal S., *Overtones and Undertones: Reading Film Music* (Berkeley, CA: University of California Press, 1994).

Brownlow, Kevin, 'Silent Films–What Was the Right Speed?', in *Early Cinema: Space, Frame, Narrative*, ed. Thomas Elsaesser and Adam Barker (London: BFI, 1990).

Cherchi Usai, Paolo, *Silent Cinema: An Introduction*, 2nd edn. (London: BFI, 2000).

Elsaesser, Thomas, 'Innocence Restored? Reading and Rereading a "Classic"', in *Fritz Lang's Metropolis: Cinematic Visions of Technology and Fear*, ed. Michael Minden and Holger Bachmann (Rochester, NY: Camden House, 2000).

Elsaesser, Thomas, *Metropolis* (London: BFI, 2000).

Eustachi, Paolo, 'Frank Strobel e Metropolis', *Colonne Sonore,* 4 August 2013. http://www.colonnesonore.net/contenutispeciali/interviste/2695-frank-strobel-a-metropolis.html.

Fischer, Tobias, 'Fifteen Questions to Frank Strobel', *Tokafi.* http://www.tokafi.com/15questions/15-questions-to-frank-strobel.

Goldmann, A. J., 'Metropolis Now', *The Wall Street Journal,* 19 February 2010. http://www.wsj.com/articles/SB10001424052748704804204575069383586138.

Gregorio, O., G. Cacciatore and D. Cappone (eds.), *Opere ascetiche. Vol. 1: Introduzione generale* (Rome: Edizioni di Storia e Letteratura, 1960).

Gunning, Tom, *The Films of Fritz Lang: Allegories of Vision and Modernity* (London: BFI, 2000).

Larsen, Peter, *Film Music,* trans. John Irons (London: Reaktion Books, 2007).

Linke, Hans-Jürgen, 'Interview mit Dirigent Frank Strobel: Von der Macht der Musik im Stummfilm', *Frankfurter Rundschau,* 28 January 2010. http://www.fr-online.de/musik/interview-mit-dirigent-frank-strobel-von-der-macht-der-musik-im-stummfilm,1473348,2824380.html.

Loew, Katharina, '14 September 1922: Schüfftan Process Reconciles Artistic Craftsmanship with Demands of Entertainment Industry', in *A New History of German Cinema,* ed. Jennifer M. Kapczinski and Michael D. Richardson (Rochester, NY: Camden House, 2012).

McCallum, Peter, 'Symphony Soars and Saws in Vivid Saga', *The Sidney Morning Herald,* 31 October 2011. http://www.smh.com.au/entertainment/music/symphony-soars-and-saws-in-vivid-saga-20111030-1mqe1.html.

McCarthy, James, 'Review of Gottfried Huppertz's Metropolis', *Limelight,* 17 August 2011. http://www.limelightmagazine.com.au/Review/267148,metropolis-original-motion-picture-score-berlin-radio-orchestra/strobel.aspx.

Miceli, Sergio, *Musica per film: Storia, Estetica, Analisi, Tipologie* (Lucca- Milan: LIM/Ricordi, 2009).

Minden, Michael and Holger Bachmann (eds.), *Fritz Lang's Metropolis: Cinematic Visions of Technology and Fear* (Rochester, NY: Camden House, 2000).

Naundorf, Karen, 'The Metropolis Mystery', *Sight and Sound,* vol. 18, no. 9 (2008), 26–9.

Patalas, Enno, 'The City of the Future – A Film of Ruins: On the Work of the Munich Film Museum', in *Fritz Lang's Metropolis: Cinematic Visions of Technology and Fear,* ed. Michael Minden and Holger Bachmann (Rochester, NY: Camden House, 2000).

Patalas, Enno, *Metropolis in-aus Trümmern* (Berlin: Dieter Bertz, 2001).

Romney, Jonathan, 'Metropolis, Fritz Lang, 145 mins (PG)', *The Independent.* http://www.independent.co.uk/arts-entertainment/films/reviews/metropolis-fritz-lang-145-mins-pg-2076981.html.

Shea, Andrea, 'Retro "Metropolis": Boston Trio Re-Scores A Sci-Fi Classic', *WBUR News,* 23 April 2010. http://www.wbur.org/2010/04/23/alloy-metropolis.

Strobel, Frank, 'Rekonstruktion und Original musik von Metropolis', *FrankStrobel.de.* http://www.frankstrobel.de/texte-von-frank-strobel/metropolis.html.

5
The Music of *The Circus*

Gillian B. Anderson

On 7 August 1993 Gillian Anderson, then a music specialist in the Library of Congress Music Division, conducted the National Symphony in a performance of the original accompaniment to Charlie Chaplin's The Circus *at Wolf Trap Farm Park for the Performing Arts in Vienna, Virginia. The Wolf Trap screening gave the audience the unique experience of being the first to see the film as it was originally presented in 1928 with its original full orchestral accompaniment and a new print. In 1968 Chaplin composed a new score for* The Circus *with Eric James,*[1] *and it is this more recent music with which Chaplin's fans are familiar. It emphasized the pathos of 'The Little Tramp' while the earlier music underscored the slapstick comedy.*

In style The Circus *is reminiscent of the two-reel films that Chaplin made for Essanay during 1915–16, which many critics regard as the purest expression of his art. In comparison to the drama of* A Woman of Paris *(1922) and the poignancy of* The Gold Rush *(1925),* The Circus *is fast-paced and loaded with the ingenious visual gags that made 'The Little Tramp' so enduring to worldwide audiences.*

Ms Anderson's account of the rediscovery of the music in the Chaplin family's archives and her reconstruction of this music indicate the painstaking labour that goes into the reconstruction of any work of art and also of the rewards that ultimately result from such work.[2]

Charlie Chaplin kept almost everything that had anything to do with his films – business records, photographs, newspaper clippings, scripts, promotional material, correspondence and music. As a professional mime, he understood well the importance of musical accompaniment, its 'hailing' function ('pay attention to this gesture, action, facial expression or body movement'), and as the following quotes make clear, music was essential to his art.

> Charlie Chaplin, not content with writing, directing and playing the chief role in *The Circus*, the comedy that goes merrily into its fourth week at the St. Francis Theater today, also supervised the music score.
>
> He places great importance on the orchestration of a moving picture, rating music an emotion-compeller second to none. Chaplin has theories about his pictures; he believes, for instance, that they should appeal to the emotions as well as to the mind. 'You cannot emphasize too much the importance of orchestra scores for pictures. I believe the score should be made as the picture is made, and not that people should spend a year making a film and then have a score written in a week.'
>
> 'It is just as if Pavlowa were to say, "I shall dance – and, oh yes – we shall have some music, too."'[3]

> 'Music is extremely important', he [Chaplin] said: 'that is why I welcome the efforts being made to provide music by mechanical systems, such as the De Forest and the Vitaphone. Mechanical music which has the quality of a symphony orchestra is much better as an accompaniment than feeble camping on a piano or the excruciating efforts of an incompetent and ill-led orchestra.'[4]

In 1992 I examined and inventoried the music in the Chaplin archives in Vevey, Switzerland and discovered the scores and orchestral parts that had been used live to accompany *The Gold Rush*, *The Circus* and *Woman of Paris* when they were first conceived or premièred during the 1920s. Much later, Chaplin wrote his own music for these moving pictures and re-released them as recorded-sound films. The scores and parts in the Chaplin archives included the later music, with which most people are familiar, as well as the earlier versions. At the time (1992), outside of the Chaplin family and members of Chaplin's business circle, no one knew of the existence of the earlier material.

Unlike his own (later) music for these films, the earliest accompaniments consisted entirely of pre-existing music chosen by Chaplin and his associates from the operatic, orchestral, popular song and dance band repertories of the 1920s. The accompaniments were put together under the supervision of various professional musician-midwives (Arthur Kay for *The Circus* and Carli Elinor for *The Gold Rush*). This meant in most cases that the music was not composed for the picture but was taken from already existing publications and arrangements. The film accompaniment was assembled by selecting music appropriate for each sequence, assigning each a rehearsal number, deciding on how much of each composition was needed, cutting out just the right number of measures, putting them in order in all the orchestral parts according to a piano/conductor score and writing any musical transitions that might be necessary.

Because Chaplin kept so many things, the parts and score for the earlier version of *The Gold Rush* were in negative photostats (instead of the positive prints that must have been made from them). The parts and scores for the earliest version of *Woman of Paris* had consecutive rehearsal numbers but not a unified piano score with cues marked in it to keep the music in synch with the film. The music had been chosen to fit certain sequences but had not yet been cut and pasted into a consecutive series. The music for *The Circus* had been numbered, cued, cut and pasted but had not been cleaned up, turned into negative photostats, the negatives touched up by painting black over the sections to be eliminated and then printed for theatrical distribution. The archive preserved parts of each stage of this process and to my knowledge was the only such collection to do so.

The first performance of the 1928 version of *The Circus* was set for a meeting of the Madison Council fund-raising group at the Library of Congress in October 1992. (The performance involved only ten instruments instead of a full orchestra.) Photocopies of the orchestral parts were mailed from Switzerland at the end of July. When I had the chance to look at them closely, I realized that they were going to be quite a challenge to restore. Usually, I have only to number all the measures in each part. I make sure thereby that there are the same number of measures in every section, where necessary cutting extra measures and reconstructing missing ones and I provide a way during rehearsals for a conductor to guide all the musicians quickly to the same spot in the music. Then I check key changes, make sure the B ♭ and A instruments are properly marked, and correct bad page turns.

In this case, however, the pre-existing printed music in each part had been pasted not over blank paper but lined music manuscript paper. This meant that occasionally when musical transitions or clefs and key signatures had to be added, the copyist had written on the music paper. However, the lines and spaces of the music paper did not ever correspond to those of the pre-existing music. It was very difficult to read and had to be changed.

In addition the music had been reduced when it was photocopied. A lot of it had to be blown back up to its original size or to a larger size so that it could be read by modern orchestral musicians (who are very particular about the state of the music from which they read).

Finally, the ledger lines, upon which the musical notes were printed, had become so faint in a number of cases that they had to be redrawn. It was painstaking, tedious work, and it had to be done quickly.

The piano part had the most notes so its restoration took forty hours of work, and it still does not look beautiful. All the pre-existing performance marks and cues had to be 'whited-out'. All the blank music manuscript paper lines had to be covered over, and the handwritten transitions on the music manuscript had to be copied over so that the lines and spaces of the pre-existing music would fit with the manuscript music.

All this restoration was necessary because the extraneous marks and the blank music ledger lines would have been extremely distracting to any modern player. There is never enough time for rehearsals, and players are almost always sight-reading their parts. Thus, a minimal number of visual distractions enhances the players' ability to concentrate on sight-reading. In many cases I was able to identify the pre-existing music used because when it was cut and pasted together, tell-tale clues were left on one or another of the orchestral parts or in the piano score. Sometimes the top of the pre-existing music with the title and composer information was left intact by the person doing the pasting. Sometimes he or she only left the bottom of the page with a publisher's serial number or copyright information. With this kind of clue I was able to identify the piece and find better copies of it so that we did not have to reconstruct the version pasted into Chaplin's original parts.

For example, halfway through rehearsal number 10, the music is almost illegible in every single orchestral part. The copyright information had been left on the bottom edge of the parts, 'Will Rossiter, Chicago, 1920'. I looked through all the Library of Congress's 1920 copyright claims by Will Rossiter and guessed that the piece I was looking for might be called 'Speed' because it accompanies Chaplin running away from the police.

When 'Speed' arrived from the Library's Landover Storage Facility, where it was stored by copyright number, sure enough, it was the right music and was in splendid shape. We substituted this music for that found illegibly in the orchestral parts. In another case, before we had found a better original, we reconstructed the music with a computer. Finally, in many cases, I had to go over bar lines with a fine-tipped pen and ruler (that's when I found out that none of the lines was really straight) and darken note stems and measure lines so that they would photocopy well.

After the first performance of *The Circus* for the Madison Council (in other words, too late to be of use in the reconstruction of ten of the parts but essential to the reconstruction of all the rest), a computer played a major role in the identification of many of the pieces used for the accompaniment of *The Circus*. Some of the most dramatic examples were provided by Russ Girsberger of the Marine Band Library. Over the years the Marine Band had accumulated a large number of early American dance band orchestrations and incidental music for small orchestra and had made them accessible via a computer system.

Out of curiosity I asked Russ what would show up under the term 'Op. 38', because all I knew about one piece of music was that it was 'From Op. 38' (information found on the lower left hand side of some of the pasted parts). A lot of composers have written an Opus 38, so I really did not expect much help and was asking Russ only out of curiosity. However, we also knew that the piece was a waltz. The Marine Band's computer system did not much like the request, but among other things it produced Edvard Grieg's 'Elegie from Op. 38, No. 6', which turned out to be the piece used by Chaplin and Kay and even the exact arrangement of it.

In another case we knew a publisher's order number, T1990, which appeared to the left of the first measure of music. Acquaintances in California had identified the composer, Tchaikovsky, but had given me the wrong title. We knew from the pieces that had already been identified that most of the pre-existing arrangements were taken from the catalogues of the American music publishers Carl Fischer, G. Schirmer, Harms, Will Rossiter, etc. Russ logged in a request for pieces by Tchaikovsky by these publishers and limited his search to music published at the beginning of the twentieth century. He found T1990, Charles J. Roberts's arrangement of Tchaikovsky's 'Humoresque. Op. 10, No. 2', which was the piece in question. Using the Marine Band computer system we also were able to identify the exact arrangement used for the excerpts from *I Pagliacci*.

But finally the computer failed. We had only a publisher's serial number from the lower left hand corner of a page, 3160-23. Russ looked for it but could not find it. In the process, however, he observed that it looked a lot like other Carl Fischer serial numbers. Carl Fischer was one of the few music publishers from the beginning of the twentieth century that was still in business. I called the publisher and asked the staff if they could identify one of their pieces just by the serial number. They said that there was a good possibility they could if the number was from the era around the turn of the century. I gave them the number, and they called back with the identification, 'R. Vollstedt, "Jolly Fellows Waltz", c NY, Carl Fischer, 1891'. The Marine Band had a copy of the arrangement, and it was the correct one.

For the ten-instrument version of the accompaniment performed in October 1992, the parts took more time to reconstruct than I had expected, so I was not able to begin synchronizing the music to the film until three weeks before the performance. I finished only one week beforehand, the closest call I have ever had – or ever care to have.[5]

One of the problems was that after I started to synchronize the music to the film, I thought that the speed of such pieces as 'Speed' was so fast as to be unplayable. The Chaplin estate had told us that the film was to be screened at 24 frames per second (fps). This was logical because the film was released very late in the 'silent' era, 1928, when recorded sound was already on its way in and a standardization of film projection speeds was well under way. However, Charlie Chaplin began production of *The Circus* in 1925 and completed it toward the end of 1927. It was released in 1928 and was the 74th film of his career, which began at the Keystone Film Co. in February 1914.

It seemed clear from the musical evidence that the speed of *The Circus* might have been slower than 24 fps.[6] For example, in the sections that were unplayable there were no cut marks in any of the parts. Clearly all of the music had been used and played. Irving Berlin's 'Blue Skies', with which the accompaniment concluded, was even mentioned in one of the New York reviews in 1928.[7]

David Francis, chief of the Motion Picture, Broadcasting and Recorded Sound Division looked at the film. He hypothesized that it should run at 20 not 24 fps, 18 per cent slower than it has been run for some thirty, forty or even sixty years.[8] This meant making a new videotape at the new speed and resynchronizing everything I had done up to that point. The unpredictable benefit of the discovery of a new speed was that audiences had time to follow the development of Chaplin's gags in a way that significantly heightened their appreciation of his humour. Therefore, not only was the music playable at the slower speed but Chaplin's art was enhanced.

The synchronization process consists of timing each scene, counting the number of beats of music in the section to which that piece of film corresponds, using an algebraic equation provided by my physicist husband, determining thereby the metronome marking, and running the videotape over and over with a metronome running until I have added enough cues in the score to keep it in synch with the film. In the process my muscles memorize the tempi; I more or less memorize the music and the movie.

'Silent' or 'mute' film scores are marked with cues for synchronization but generally only at the beginning of major sections or over chords that require exact synchronization. *The Circus* compilation is liberally cued but my copy is even more so. For example, at rehearsal No. 2 one finds 'T. The Circus'. This means that when that title appears on the screen, the music for rehearsal No. 2 begins. I added many additional cues to make sure that when the heroine breaks through the paper in a hoop 20 measures later, the action can occur on exactly the right chord. It is clearly meant to be that way.

The music for *The Circus* displays not only Kay and Chaplin's wide knowledge of music but their extraordinarily good judgement about what makes an apt marriage of music and image. In one scene, 'The Little Tramp' finds himself locked in a lion's cage. The lion goes up and sniffs him but walks away bored. After the heroine, Merna Kennedy, unlocks the cage door, Chaplin, full of bravado, approaches the lion to the music of the 'Toreador Song' from *Carmen*. When the lion roars, Chaplin tears out of the cage. He is hardly a brave toreador, the lion hardly a bull. The music enhances the humour of the scene.

When Chaplin climbs onto a high wire, supported by a belt with a wire attached, and goes through the routine of a tightrope walker, music performs a similar function. The belt comes undone without Chaplin realizing it just at the same moment that a bunch of monkeys climb onto the tightrope with him. Full of confidence that his wire and belt are secure, he does a dance on the wire, to the accompaniment of a 1920s dance band arrangement of James P. Johnston's 'Charleston'. The music emphasizes his confidence, foolhardiness and the impending disaster. In other places the music underscores the story of the sad clown. The movie opens with music from Leoncavallo's *I Pagliacci*. We hear Gounod's 'Funeral March of a Marionette', Victor Herbert's 'Punchinello', Arthur Kay's 'A Funny Story', and Lessing and Monaco's 'Oh! You Circus Days', all on the same theme.

Chaplin and Kay employed the popular music of their day as well. The accompaniment for *The Circus* uses Nacio Herb Brown's 'The Sneak' (1922), Ray Henderson's 'Just a Memory' (1927), Rudolf Friml's 'L'amour toujours l'amour' (1922), Victor Herbert's 'Some Day' from *Her Regiment* (1917) and 'For I'm Falling in Love with Someone' from *Naughty Marietta* (1910), a 1926 arrangement of Scott Joplin's 'Maple Leaf Rag', DeSylva, Brown and Henderson's 'Lucky Day' (1926), Robert Katscher's 'When Day is Done' (1926) and Irving Berlin's 'Blue Skies' (1927).

Most people today are not familiar with these pieces, but in 1928 the audiences would have known not only the tunes but also the words. When Charlie is shown asleep in a wheelbarrow after his harrowing escape from the police, they would have recognized two measures of 'Please Let Me Sleep'. When the tightrope walker and Merna Kennedy first meet, they would have recognized the music as 'For I'm Falling in Love with Someone'. When Charlie overhears Merna telling a fortune-teller that she has just fallen in love, thinking that she is speaking of him, Charlie is overjoyed. The audiences would have recognized the music as Henderson's 'Lucky Day'. At the end of the show, when the circus train has departed, leaving Chaplin alone with a star on a piece of paper (Merna had jumped through it in the first scene), everyone would have known the first line, 'Blue skies, smiling at me, nothing but blue skies do I see.' Many of us still know this line well enough to feel the music underlining the pathos of the finale.

'The Sneak', described as 'the greatest novelty foxtrot song in a decade', accompanies the scene in which a thief is nabbed by a policeman. Nacio Herb Brown, its composer, was the bandleader at a club in Hollywood that Chaplin frequented.[9] Chaplin probably knew all the songs from such places, but his and Kay's choices enhanced both the slapstick comedy in *The Circus* and the pathos of 'The Little Tramp'. They added a musical commentary to the picture.

The ultimate test of the success of a wedding between image and sound is the actual screening. In 1928 the reviews for Chaplin's 'low-brow comedy for highbrows'[10] were raves:

> *CIRCUS* HOWLING AT THE MARK STRAND. CHARLIE ONCE AGAIN PROVES A REVELATION. Charles Spencer Chaplin's *Circus* is a screaming delight from fade-in to fadeout. It is a howling, hearty, happy, slightly slapstick production wherein the inimitable Charlie gets you more often by a laugh than by a tear. One is kept so constantly in a state of grin, giggle and guffaw at this glorious picturization of the [...] tramp who attains tent show fame, but not love, that even in the moments of pathos – which don't number nearly as many as did several previous Chaplin vehicles – one doesn't weep freely. Behind each tear there are at least a dozen laughs.[11]
>
> The hardest-boiled crowd in town went to the midnight opening on Friday and laughed off all its mascara.[12]

Sixty-five years later, modern audiences responded the same way. There have been at least seventeen[13] performances of *The Circus* with its original accompaniment. One review of a performance at Chaplin's own studio in Hollywood in March, 1993 stated:

> The unblemished images of Chaplin's masterful gags and astonishing physical prowess had the audience whooping with delight. When, at the film's climax, the tramp performs a high-wire act as a horde of monkeys runs down the wire and attacks him, the audience gasped as loudly as I've ever heard in a theater. [...] The musicians and the conductor performed with admirable fortitude for the film's 90 minutes, ceaselessly synchronizing the film without the use of click tracks. [...] When, at last the lights went on, the viewers lingered in their seats, reluctant to leave behind the sound stage, the tramp, and the timeless comedy he created there seventy-five years ago.

The Washington Post said of the 7 August 1993 Wolf Trap performance:

> The juxtaposition of the music with the action of the film is brilliant. For example, early in *The Circus*, as Chaplin is seen sprinting from a pursuing policeman, the music keeps a pulse matching the movement exquisitely. Various percussive accents highlight abrupt maneuvers, and touching musical satire is implied at choice moments. Anderson must be commended, not only on a solid performance but also for obtaining this valuable music. The feat was matched only by the stamina of the orchestra, which was required to play continuously for 90 minutes.[14]

One of the greatest pleasures for me has been to hear the response of young children to this picture. Their giggles and laughter, an octave above that of the adults, were frequent, and when 'The Little Tramp' ran into the lion's cage, one child yelled out, 'Oh, no!' at the top of his lungs, obviously completely engrossed in this motion picture. It is simply a hilarious movie for people of all ages.

Rarely has a general audience had the treat of seeing a gorgeous new print with a wonderful accompaniment played superbly by the National Symphony. The response was unanimous. After laughing hysterically for ninety minutes, the Wolf Trap audience applauded the genius of 'The Little Tramp', which had been released for their rediscovery and entertainment by the slower speed, a new print and the glorious performance of its first orchestral score.

Curiosity was my motivation for the detective work and the myriad tiny, tedious operations detailed in the above description. I was curious about how the compiled score for *The Circus* functioned and whether it played a role in the film's reception. Would it work with the picture? What might it reveal about Chaplin or his collaborator Arthur Kay? How did it compare to the later score composed by Chaplin with the help of Eric James [Barker]? What would the music reveal about the movie itself? Together with my other restored or reconstructed scores, might it contribute to some overall pattern that might not have been taken into consideration by musicologists or cinema studies specialists? Might it give us any historical insights that might not be available in any other way than by reconstructing and performing the work?

It is still (shockingly) true that only a handful of mute films are available on DVD with their original orchestral scores,[15] and fewer still are synchronized as indicated in the scores.[16] So we have thirty years of early cinema history based almost entirely on the images alone or images with today's music, even though, originally, there was a vast music-making apparatus that accompanied the picture shows. Until 1929 many involved in the industry (Chaplin was only one of them) considered moving pictures to be a subset of pantomime, a form for which a synchronized musical accompaniment was considered utterly essential.[17] The musical accompaniments for good or ill did (and do still) affect the reception of 'mute' pictures, but sadly in most cases we have no way to compare the reception history preserved in newspaper and magazine accounts with an actual moving picture with its original accompaniment. Half the history is missing because the actual historical artefact, a moving picture with its original accompaniment, cannot be viewed. However, particularly in the case of Chaplin, Griffith and Fairbanks who attempted to control the musical presentation of their moving pictures, the music was part of the directors' intentions.

Of course we cannot put ourselves back into the heads or ears of the audiences, composers or musicians at the beginning of the twentieth century. Too much has changed. We can only understand the works of one hundred years ago with today's ears, eyes and brains. However, *The Circus* is a good case study of the sorts of insights one can gain by putting a movie back together with its orphaned accompaniment.

Clearly the first score worked and still does. The music caught the tempo of the actions. 'Speed' aptly went with the chase scene and 'Blue Skies' deepened the pathos of the ending. Some of the humour in the picture was even carried further by the music that was chosen. (The 'Toreador' music for the lion's cage.) Clearly Chaplin considered the motion picture a subset of pantomime and music an essential element. Clearly the music was compiled and synchronized with the image (contrary to the many who assert that synchronized sound only arrived with the talking picture). The Chaplin/Kay compilation was clearly successful. The reception history for the Chaplin/James score is not so clear. On the one hand one finds:

> For the opening of the reissue version of *The Circus*, Chaplin replaced *I Pagliacci* with a ballad entitled 'Swing, Little Girl'. Chaplin sang the expressive lyrical ballad himself over remade opening credits. In place of Bizet's 'Toreador Song' from *Carmen*, Chaplin underscored the lion sequence with a simple vamp complemented by an ominous lion theme (to mimic the lion's yawn) interspersed with a dainty string phrase (to emphasise the Tramp's caution). For the sequence where the high-wire walking Tramp encounters the bites and paws of the monkeys, Chaplin had initially employed Minot's 'Hurry No. 26' to mirror the action. For the reissue, Chaplin chose a slow, pleasant waltz as an effective counterpoint to the chaotic action. Finally, for the ending of the film, instead of Berlin's 'Blue Skies', Chaplin began the closing sequence with a dramatic fanfare (in a minor key) as the circus wagons pull away leaving the Tramp alone in the centre ring. Following the fanfare, Chaplin employs a ballad (still in the minor key) which builds its sad melody to a crescendo as the Tramp walks down his lonely road.
>
> On the whole, Chaplin's 1970 reissue score conveys a more effective marriage of image and sound. Like so much of his music, the best word to describe Chaplin's 1970 reissue score of *The Circus* is quite simply that it is Chaplinesque and indeed some of the finest music Chaplin ever composed.[18]

On the other hand one finds this criticism of his music in general:

> in connection with [...] *Monsieur Verdoux* and *City Lights*. Funny as these pictures are, they leave an audience with the feeling that the humor constantly crosses over into areas of profound thought and feeling where all simplicities are complex and all complexities are apparently insoluble. But this is not true of Chaplin's music. It is quite bald and literal. Chaplin seems to find in music only its gestural meaning and its surface emotion. Hence it seldom rises above cue catching and sentimentality, and except for the elegance of performance that nearly all music gets on sound stages nowadays, there is little to distinguish a Chaplin score from what theater organists used to provide for the silent comedies. [...] The boulevard music in

> *Verdoux*, for instance, would be appropriate in any film with a Parisian setting, and the train music is just as generalized. The only kind of wit and humor that Chaplin recognizes is the quotation of familiar themes in incongruous circumstances. [...] But this is a literary, not musical, humor [...] Stylistic parody, on the order of the 'potted palm' music [...] has been slightly cultivated in Hollywood, though not by Chaplin, who ought to be its past master. Two instances, which I have cited frequently before, come again to mind: Friedhofer's parody of the concerto grosso style in *The Bishop's Wife*, and Raksin's pseudo-Handelian music for the amusing scene in the King's antechamber in *Forever Amber*. Irony of this kind is at the very core of Chaplin's picture making, but it never finds its way into his music. I would attribute its absence to the naïveté of his musical tastes. He seems to be unaware of the cultivation of wit by contemporary composers, the wit of the misplaced accent, the unsymmetrical phrase, the comically distorted harmony, or the purposeful misuse of an instrument. Such devices were not, to be sure, invented by our contemporaries. [...] And it is in this tradition that Chaplin ought to be finding his own musical methodology. It is ironic that Eisbrenner, in *Wonderful Times*, should have stumbled accidentally into procedures that Chaplin seems not even yet to be aware of.[19]

This is a criticism that could be made equally of both scores for *The Circus* and perhaps most 'mute' film scores as well. Obviously, a compiled score can be successful without achieving the standard set by Lawrence Morton. However, his criticism could not be made of the score for *Modern Times* (for which Chaplin's midwife was David Raksin). One might conjecture that Chaplin carried the 'mute' film score conventions into the recorded-sound era. Only arguments with David Raksin (he was fired more than once) prevented this outlook from controlling the score to *Modern Times*.[20] For sure, the composer-midwife Chaplin used affected the quality of a score's outcome. For *The Circus* Arthur Kay was an extremely competent compiler with very good taste. In my opinion the Chaplin/Elinor score for *The Gold Rush*, by comparison, was not as consistently effective.[21]

A performance of the Chaplin/Kay score for *The Circus* together with numerous other scores of the same period[22] convinced me that these scores *were* synchronized with the moving pictures. Stop watches and metronomes were used to select exactly the right number of measures of music to fill the time necessary for a scene or part of a scene. As shown in reviews,[23] often in performance the intended synchronization was realized but as often (as evidenced by the Chaplin quote at the beginning of this chapter) it was not. With the talking picture the synchronization was mechanically controlled by the recording process. It was standardized. The difference between the synchronized sound for 'mute' and talking pictures was the difference between live and recorded sound, *not* non-synchronous and synchronous sound. Only by performing, restoring or reconstructing these scores was I brought to this conclusion.

As a professional mime, Charlie Chaplin knew well the essential value of highly synchronized music. As a composer he depended on the aid of various trained musicians to bring his musical ideas to fruition. After performing three of the scores for his films, I would say that some of his assistants were more successful than others. This too is something I could only have concluded after performing the works and seeing how they functioned with an audience.

The Circus is not long, 71 minutes. After its première at the Mark Strand Theatre in New York where it had a brief prologue,[24] it often appeared at movie theatres with a live vaudeville programme[25] that had an orchestra,[26] thus a supportive musical apparatus. However, as indicated by Chaplin's comments at the outset of this chapter, the film also would have been programmed into theatres where the musicians were not adequate. For this reason he would have welcomed the advent of the mechanically reproduced music track that arrived with the talking picture. It took the musical performance and the synchronization out of the hands of live theatre musicians.

Having performed the Chaplin/Kay compilation for *The Circus*, I realized that it required the execution of good musicians to be successfully synchronized. I came to realize that in the mute film era many regarded the motion picture as a form of pantomime for which synchronized music was essential. I came to understand that a professional mime like Chaplin would have preferred a medium where the music and the synchronization were more rigorously controlled, mechanically, and it made even more sense that eventually he would compose his own musical accompaniments, replacing the compiled scores that had been used in the 1920s. I don't believe I would have realized any of this without the experience of actually performing the original score for so many of the moving pictures of the 'mute' era.

When I was first asked to submit a chapter to this volume, I said no. I detest new scores that are written willy-nilly without any knowledge of or respect for the moving pictures they accompany, the drama being presented, the epoch in which they were created or the conventions upon which they and their musical accompaniments were based.

I dislike scores that scream, 'Listen to what a brilliant performer or composer I am', where the audio completely overwhelms the video, reducing the film to an excuse for a concert. I do not like cheap performers who clearly use already made movies in order to save the cost of making a new one. Nor am I crazy about those who, in order to save the money that would be necessary to hire an orchestra, have a pianist bang away, improvising endlessly without any sense of musical direction. However, I have submitted this chapter because I am now convinced that the existence of so many accompaniments for 'mute' films is a phenomenon that is worth studying and I want to make my position clear within this context. A lack of attention to the original accompaniments for 'mute' films has led to some egregious historical errors and an underestimation of the power and function of music to the moving image. Every time a new score or an improvisation is used for a film that had an old one, even just a cue sheet, it is an opportunity lost to fill in an historical blank. As can be seen from its description, the work that went into the restoration of the Chaplin/Kay compilation score for *The Circus* is extremely tedious, not at all sexy, but the end result – to understand the impact of the images with their original orchestral accompaniments, to see a sleeping giant come to life – is very rewarding indeed.

Notes

1. Eric James, Making Music with Charlie Chaplin: An Autobiography. Filmmakers Series No. 71 (Lanham, MD: Scarecrow Press, 2000), 63–77. See also Kenneth Schuyler Lynn, *Charlie Chaplin and His Times* (New York: Simon & Schuster, 1997) and Jeffrey Vance, Chaplin: Genius of the Cinema (New York: Harry N. Abrams, 2003).
2. A version of this article first appeared in the Library of Congress *Information Bulletin*, vol. 152, no. 17 (30 September 1993), 341–9.
3. 'Fourth Week for Chaplin in *The Circus* at St. Francis. Music Score Supervised by Film Star. Funster Chose Supporting Actors, Wrote Story and Directed Movie', *San Francisco Chronicle*, 19 May 1928, CCC7.
4. L'Estrange Fawcett, *Films: Facts and Forecasts* (London: Geoffrey Bles, 1930), 153.
5. See Gillian B. Anderson, 'Synchronized Music: The Influence of Pantomime on Moving Pictures', *Music and the Moving Image*, vol. 8, no. 3 (Fall 2015), forthcoming, for proof that music in the mute film era was synchronized.
6. Although speed is subjective and today the tendency to produce pear-shaped tones has slowed down the velocity of music that used to be played lightly on gut strings.
7. Mordaunt Hall, 'The Screen. Chaplin of Hollywood. *The Circus*', *New York Times*, 9 January 1928, 29.
8. Documents at the Chaplin Estate, however, confirmed that the speed of the film had been 24 frames per second.
9. I am indebted to Lance Bowling of Cambria Records for this information.

10. 'Close-Ups', *Boston Herald*, 21 September 1927, 21.
11. Irene Thierer, *New York Daily News*, 9 June 1928, quoted in 'The Circus United Artists Strand', *The Film Daily*, 11 January 1928, 7. Also quoted in 'Newspaper Opinion on New Pictures. *The Circus* – United Artists, Mark Strand Theatre, New York City', *Motion Picture News*, 4 February 1928, 400.
12. Bland Johanson, *New York Daily Mirror* quoted in 'The Circus United Artists Strand', *The Film Daily*, 11 January 1928, 7.
13. Library of Congress, National Symphony at Wolf Trap Farm, Palm Beach, FLA, San Diego Symphony, Milwaukee Chamber Orchestra, Chaplin Studio, MIT, University of Wisconsin, Michigan Theater, Hunter College, Metropolitan Museum of Art, Teatro Comunale Orchestra, Bologna, Orquestra classica da Porto, Lisbon, the Orchestra Symphonique d'Europe, Paris, the Orchestra sinfonica dell'Emilia Romagna, Parma, Malmö Symphony Orchestra, members of the Strasbourg Symphony, Strasbourg.
14. Bob Waters, 'National Symphony Orchestra', *The Washington Post*, 7 August 1993.
15. Especially if one does not count scores created by a synthesizer instead of real acoustic instruments, *Wings* and *The Thief of Bagdad* for example.
16. *The Birth of a Nation* for example is available on DVD with the original Breil score but bears no relation whatever to the cues in the score.
17. See Anderson, 'Synchronized Music: The Influence of Pantomime on Moving Pictures'.
18. Jeffrey Vance, '*The Circus*: A Chaplin Masterpiece', *Film History*, vol. 8, no. 2, The 1950s and beyond (1996), 186–208.
19. Lawrence Morton, 'Film Music of the Quarter', *The Quarterly of Film Radio and Television*, vol. 6, no. 1 (Autumn, 1951), 69–72.
20. David Raksin, 'Life with Charlie', in *Wonderful Inventions: Motion Pictures, Broadcasting, and Recorded Sound at the Library of Congress*, ed. Iris Newsom (Washington, DC: Library of Congress, 1985), 158–71. I performed the Chaplin/Raksin masterpiece live with the film and the FVG Orchestra of Udine, IT.
21. I performed the Chaplin/Elinor score for *The Gold Rush* at the Kennedy Center in Washington, DC.
22. See http://www.gilliananderson.it.
23. See for example Gillian B. Anderson, 'D. W. Griffith's *Intolerance*: Revisiting a Reconstructed Text', *Film History*, vol. 25, no. 3 (2013), 57–89 and Anderson, 'Synchronized Music: The Influence of Pantomime on Moving Pictures'.
24. 'Prologue Flash on Strand Stage; Simple, but Highly Effective', *The Film Daily*, 15 January 1928, 4: 'With the Charlie Chaplin feature, *The Circus*, carrying practically the entire performance, there was little time for a stage presentation. It was confined to a very brief prologue which nicely set the atmosphere for the picture. The curtains parted on a circus flash which had been announced by a "barker" through a phonograph. The set showed a circus ring with the master of ceremonies in his red coat and long whip directing the aggregation of trained wild animals. These were immense papier-mache constructions which were ingeniously animated and performed antics as the leader snapped the whip. It was just sufficient to set the audience nicely for the circus atmosphere of the picture.'
25. 'Loew's State', *Boston Herald*, 25 March 1928, 73; 'Charlie Chaplin Star at the State. Creates Lots of Fun in *The Circus* – Novel Musical Features in the

Stage Show', *Boston Globe*, 27 March 1927, 12; 'Charlie Chaplin's *Circus* Film at the Orpheum', *Boston Globe*, 8 April 1928, A52; *San Francisco Chronicle*, 13 May 1928, 87; 'Manager George E. Brown of the Carolina theater said that he felt "highly gratified" in being able to present Charlie Chaplin's first comedy in two and a half years', *Seattle Daily Times*, 8 March 1928, 10; 'Coronado Theatre', *Rockford Republic* (Rockford, IL), 19 March 1928, 10; K.T.K., 'Asbestos. Saenger. *The Circus*', *The Times-Picayune*, 6 February 1928, 6; 'Charlie Chaplin in *The Circus*', *Trenton Evening Times*, 25 February 1928, 13. In England this type of programme was later called cine-variety. James, *Making Music with Charlie Chaplin*, 15.

26. '*The Circus* – Columbia', *Seattle Daily Times*, 29 February 1928, 18: 'The Columbia Theatre is "packing them in" with *The Circus*. [...] Francesco Longo and the Columbia All-Artist Orchestra accompany the picture with a score arranged by Longo and are featured in concert'; 'Carolina Theatre', *Greensboro Record* (NC), 26 February 1928, 27: 'The Orchestra. Monday the Carolina introduces its new Symphony orchestra, which takes rank as the finest concert orchestra in the Carolinas. Louis J. Valle, conducting'; 'Coronado Theatre', *Rockford Republic* (Rockford, IL), 19 March 1928, 10: 'Coronado Symphony Orchestra Direction Maurice Hazard. Ray Turner at the Barton Grande Console. Offering "A Little of This and Little of That"'; 'Anniversary Show at New Saenger Has Chaplin Film. "Flying High" is stage show with Jerrie, Accordionist', *The Times-Picayune*, 5 February 1928, 36: 'Castro Carazo and the orchestra will offer as the anniversary overture'; K.T.K., 'Asbestos. Saenger. *The Circus*', *The Times-Picayune*, 6 February 1928, 6: 'Helen Yorke and S. Giacomo sing a "Birthday Greeting" medley of airs, with the orchestra under Mr. Carazo. Mr. Hammond takes his audience for an interesting "Trip Through the Organ"'; 'Coronado', *Daily Register Gazette* (Rockford, IL), 22 March 1928, 22: 'On the stage Eddie Moran, Cole & Wray, Smith & Hadley, Symphony Orch. Direction Maurice Hazard'; 'Loew's Lexington', *New York Times*, 11 March 1928, 112: 'Charlie Chaplin in *The Circus* Concert Orchestra and Organ'.

Bibliography

Anderson, Gillian B., 'D. W. Griffith's Intolerance: Revisiting a Reconstructed Text', *Film History*, vol. 25, no. 3 (2013), 57–89.

Anderson, Gillian B., 'Synchronized Music: The Influence of Pantomime on Moving Pictures', *Music and the Moving Image*, vol. 8, no. 3 (Fall 2015), 3–39.

'Anniversary Show at New Saenger Has Chaplin Film. "Flying High" is stage show with Jerrie, Accordionist', *The Times-Picayune*, 5 February 1928.

'Carolina Theatre', *Greensboro Record* (NC), 26 February 1928.

'Charlie Chaplin in The Circus', *Trenton Evening Times*, 25 February 1928.

'Charlie Chaplin Star at the State. Creates Lots of Fun in The Circus – Novel Musical Features in the Stage Show', *Boston Globe*, 27 March 1927.

'Charlie Chaplin's Circus Film at the Orpheum', *Boston Globe*, 8 April 1928.

'Close-Ups', *Boston Herald*, 21 September 1927.

'Coronado', *Daily Register Gazette* (Rockford, IL), 22 March 1928.

'Coronado Theatre', *Rockford Republic* (Rockford, IL), 19 March 1928.

Fawcett, L'Estrange, *Films: Facts and Forecasts* (London: Geoffrey Bles, 1930).
'Fourth Week for Chaplin in The Circus at St. Francis. Music Score Supervised by Film Star. Funster Chose Supporting Actors, Wrote Story and Directed Movie', *San Francisco Chronicle*, 19 May 1928.
James, Eric, *Making Music with Charlie Chaplin: An Autobiography*, Filmmakers Series No. 71 (Lanham, MD: Scarecrow Press, 2000).
K.T.K., 'Asbestos. Saenger. The Circus', *The Times-Picayune*, 6 February 1928.
'Loew's Lexington', *New York Times*, 11 March 1928.
'Loew's State', *Boston Herald*, 25 March 1928.
Lynn, Kenneth Schuyler, *Charlie Chaplin and His Times* (New York: Simon & Schuster, 1997).
'Manager George E. Brown of the Carolina theater said that he felt "highly gratified" in being able to present Charlie Chaplin's first comedy in two and a half years', *Seattle Daily Times*, 8 March 1928.
Mordaunt Hall, 'The Screen. Chaplin of Hollywood. The Circus', *The New York Times*, 9 January 1928.
Morton, Lawrence, 'Film Music of the Quarter', *The Quarterly of Film Radio and Television*, vol. 6, no. 1 (Autumn,1951).
'Newspaper Opinion on New Pictures. The Circus – United Artists, Mark Strand Theatre, New York City', *Motion Picture News*, 4 February 1928.
'Prologue Flash on Strand Stage; Simple, but Highly Effective', *The Film Daily*, 15 January1928.
Raksin, David, 'Life with Charlie', in *Wonderful Inventions: Motion Pictures, Broadcasting, and Recorded Sound at the Library of Congress*, ed. Iris Newsom (Washington, DC: Library of Congress, 1985).
'The Circus – Columbia', *Seattle Daily Times*, 29 February 1928.
'The Circus United Artists Strand', *The Film Daily*, 11 January 1928.
Vance, Jeffrey, *Chaplin: Genius of the Cinema* (New York: Harry N. Abrams, 2003).
Vance, Jeffrey, 'The Circus: A Chaplin Masterpiece', *Film History*, vol. 8, no. 2 (1996), 186–208.
Waters, Bob, 'National Symphony Orchestra', *The Washington Post*, 7 August 1993.

6

Cowboys, Beggars and the 'Deep Ellum Blues': Playing Authentic to Silent Films

Michael Hammond

It was on a cool October evening in 2007 under outdoor heaters outside a café in Pordenone, Italy, that composer and writer Neil Brand suggested to me that we might collaborate on playing to a silent film. As a film scholar and musician I was intrigued by the thought of putting these two separate spheres together. His suggestion was that we play to either W. S. Hart's 1916 masterpiece *Hell's Hinges* or his 1921 film *White Oak*. His reason for approaching me is that I play in a jugband/skiffle group called The Dodge Brothers, along with well-known UK film critic Dr Mark Kermode. Neil had heard us play and was familiar with our material (and our limitations) and thought that together we could bring something to the Hart films. What attracted Neil to us, apart from the guaranteed national publicity that Mark brings, was that we draw much of our repertoire from what Greil Marcus, paraphrasing the poet Kenneth Roxroth's 'old free America', has called the music of the 'old weird America', or what I would call the 'old weird American songbook'.[1] These are folk and popular songs which range from the old timey proto-country songs of the Carter Family, Jimmie Rodgers, Dock Boggs and Charlie Poole to the country blues of Furry Lewis, Charley Patton, Bukka White, Geechie Wiley and L. V. Thomas. Much of this music dates from before the First World War but was recorded by these artists in the 1920s and 1930s. The Hart films lend themselves to the kind of dark morality plays that these songs evoke, murder ballads, gambling songs, train wrecks and work songs. Also Neil knew that our approach was more geared towards an authenticity of process rather than attempting to imitate these players. This means that we work in the same way itinerant musicians of the time did, which was to play songs in our own way, although within their broad performance tradition. (We jump around, tell jokes during songs, tap dance, play wine bottles and stand up on the double bass.) Neil thought that our style would be a good match for the Hart films and later, after a successful engagement at the Barbican Centre in London with *White Oak*, brought to our attention William Wellman's 1927 *Beggars of Life*.

Since doing the first performance at the British Silent Film Festival in 2009, we have been invited to perform at festivals across the UK at Bradford, Shetland, the Isle of Man and most recently the pop music festival at Glastonbury, as well as the National Film Theatre in London. The reviews for our performances have been uniformly positive and have highlighted our brand of playing as bringing an 'authenticity' to the films. Given that Neil and us have consciously aimed to give these performances a patina of historical verity I want to use this experience as a basis to outline some thoughts on what the implications of having a sense of the historical contexts of the films and their original performance conditions and approaches has for the live performance to silent cinema in the early twenty-first century.

'A small film theatre, somewhere in Texas'

Our first performance was for Hart's *White Oa*k at the Barbican Centre in London. We had decided on that film primarily because of the availability of a good print and the fact that the Western theme leant itself well to what the Barbican described as our 'unique blend of American retro'. In introducing the film I asked the audience to imagine they were seeing the film in a small motion picture theatre in a town somewhere in Texas. I did that partly because the resonance of invoking Texas in the same breath as 'western' gave the performance an air of showmanship. But there was a certain historical accuracy to my request because I knew that such a theatre would indeed have had a small ensemble, like ours, distinct from the orchestras in larger theatres in cities. And as we were about to demonstrate, those kind of small ensembles had the potential to employ a wider latitude of musical interpretation than would have been intended by Hart perhaps, and most certainly by musical directors like Hugo Reisenfeld who oversaw the large orchestras in urban theatres such as the Rivoli in New York City throughout the silent period.

Flexibility with the accompaniments was the norm for most cinemagoers' experience in the silent period. Though the late teens and early 1920s was a period where the industry was moving in fits and starts towards standardizing film music, along this winding road there were numerous U-turns, sidetracks and dead-ends where playful, if not insolent, engagements with the film on the screen by musicians and audiences took place. Recent scholarship on early film sound points to broad developments which made sound central to the establishment of commercial film styles and the rise of the dominant studio system. The important development in the first two decades of cinema was the use of sound to 'silence' the audience.[2] The heterogeneous environments of this early period ranged from carnival tents and vaudeville to scientific lectures and store-front nickelodeons. Each of these had different performance traditions that extended beyond the stage to include audience behaviour. In his comprehensive history of silent film music Rick Altman describes the cacophony of the soundscape of an early nickelodeon which involved the barker at the storefront competing with the 'ballyhoo' music, usually either a player piano or gramophone, that played the popular tunes of the day into the street to attract passers-by. While this was going on at the front, the pianist inside the nickelodeon 'battled audience remarks from time to time interrupted by an operator's announcement, an illustrated song, or mechanical music' in the form of a player piano or gramophone.[3] By the late teens and early 1920s the emphasis on silencing the audience and taming this chaotic environment had somewhat given way to strategies that sought to standardize the experience. For example, placing the musicians in front or near the screen enabled the sound accompaniment to come from one spatial source and had the effect of connecting the music to the screen. This technique was within the tradition of stage music and also that of the 'combinations' such as Al W. Martin's 1898 production of *Uncle Tom's Cabin* or Scott Marble's 1897 *The Great Train Robbery*, both of which featured dancing and singing acts built into the plays.[4] Yet, while these influential forms shaped the way that silent film accompaniment developed, they themselves had been based on a variety programme of shorter acts rather than an unbroken feature-length narrative. In these performances the audience's attention was directed to a number of different attractions and the early film audiences and musicians were accustomed to this type of spectatorship. Although film music became a way of regulating and guiding audience experiences while offering a means of standardizing that experience for an increasingly national and global industry, the move to a well-trained and silent gazing and listening audience was neither uniform nor fully complete (if ever) until the late 1920s. The picture that has emerged of these early decades is one of a chaotic and diverse soundscape and a developing industry and art-form shaped by attempts by the industry to standardize film sound and music.

Controlling unruly audiences as well as competing sound and music sources was an important motivation in attempts by the burgeoning studios to regulate the cinema screens and the audience. Binding the music specifically with the narrative of the film required more attentive listening as well as viewing. Standardizing the kind of music that would be played to specific films was begun in the 1910s and by the 1920s scores and cue sheets put together by the production companies to accompany new releases had become more common. Increasing centralization accompanied by choice of film music was a form of regulation through 'education' and it is here that high-brow classical music was intermixed with recognizable folk and popular tunes. What evidence exists of this lies in the trade press, surviving scores and cue sheets, much of it recorded by writers and composers for whom standardizing accompaniment and choice of music was the main driver. However, the possibility, and the evidence, that musicians might indulge in changing and shaping accompaniments for a specific local audience is apparent in the writing of advisers in the trade press. Altman cites Clarence E. Sinn of *Moving Picture World* who in 1910 advised against live accompaniment that played to specific local audiences and recommended a focus on the intentions of the film producer who 'takes great pains to convey certain impressions and preserve a certain atmosphere [...] so the audience may receive his story in the same spirit in which it is told'.[5] His advice provides us with evidence of the lively atmospheres created by audiences and musicians. Seventeen years later, Harold B Franklin, President of the West Coast Theatre Inc. in his 1927 book *Motion Picture Theatre Management* outlined the degree to which Sinn's model for musical accompaniment had been achieved by large urban theatres such as the Criterion and the Rivoli in New York City:

> The fine theatre boasts of very extensive music libraries, which include works of a wide range, where the gems of classic composers may be found side by side with the offerings of popular contemporary writers. The classics have greatest value because they have endured, and because in the final analysis they have the greatest heart appeal. Jazz has its place on the musical programme when the occasion requires it.[6]

Franklin's claim also demonstrates the degree to which popular forms had been integrated into the standard norms for accompaniment by the late 1920s. While encouraging the classical motifs and attention to the demands of the narrative, he also shows a keen sense of the need for variety in selections that appeal to audiences.

The inclusion of jazz no doubt reflects the popularity of the form and its incorporation was one solution to the problem that musicians and the industry as a whole faced with the day in and day out playing of music to films on continuous programmes. The variety required was considerable as musicians could not rely on reusing the same classical motifs, or the same folk songs. In any given week most theatres changed programmes of eight to ten films at least twice and sometimes three times a week. Playing to 25–30 different films in a week meant that musicians could not rely on the same pieces of music day in and day out. The popular music industry offered a partial solution in the rapid turnover of popular songs being published as sheet music every week and recorded on gramophone records, and later in the 1920s heard on radio. These renewable sources gave the musicians a means of providing variety along with some degree of certainty that the music would connect with audiences. Franklin's approval of jazz, while somewhat reluctant, provides a glimpse of the atmosphere of specific instances where musicians relied on more improvisatory forms than either classical or well-known folk or traditional songs.

The recommendations for the appropriate musical accompaniment to films to be found in the trade press such as Sinn's in *Moving Picture World*, in film exhibitors' manuals such as Franklin's or the debates about appropriate music in the journal *The American Organist* throughout the 1910s and 1920s, all indicate that there was a divergent range of experiences in the cinemas across the United States. For me there is inspiration in the rich diversity of music evident by virtue of what these contemporary critics and advisers are condemning. When we invoke a 'small theatre in a Texas town' to set the atmosphere for a modern audience it is to encourage a sense of 'locality' and to raise these historical experiences, these impudent revenants, from the secret history that the standardizing discourse of regulation suppressed.

In part I have in mind Mary Carbine's groundbreaking and evocative study of the film accompaniment in Chicago's African American theatres in the 1910s and 1920s. Carbine outlined a similar regulating discourse in orchestra leader Dave Peyton's column in the black newspaper *The Chicago Defender* entitled 'The Musical Bunch'. 'Peyton's columns exhibited a double strategy for enhancing black opportunity and cultural identity: on one hand pursuing the legitimacy of European music and respectable behavior while, on the other, guarding black musical invention from appropriation and exploitation by whites.'[7] Peyton wrote in one column that black musicians should guard against giving away the secrets of their techniques to white musicians, while in another noted that 'certain so-called jazz artists [are] getting away from the score and the unqualified leader not being aware that the jazz artist is ruining the composition'.[8] By pointing to the discourse of 'appropriate' music Carbine is able to outline a detailed picture of the kind of atmosphere in black cinemas in Chicago that illuminates the various uses of music on the programme ranging from specialist live performing acts to the cinema orchestra that at times might interrupt and hijack the films on the screen, which were predominantly Hollywood products, as well as supporting the film along the lines that Sinn and Franklin suggest.

The urban black community in Chicago at this time was somewhat unique in its position as a major destination in the 'Great Migration' of African Americans to the northern cities from roughly 1915 to 1970. It was a potent environment for the development of blues and jazz styles and innovations, traditions brought from all parts of the South. Carbine notes that Peyton's broad tendency to discourage jazz improvisation did have some tolerance and it was allowable, just as Franklin did later, in certain circumstances such as comedy films. However he disliked 'the awful, low-down, so-called blues' and wrote that they 'should be eliminated entirely from the pit'.[9] Those blues he refers to are the music of artists like Ma Rainey and Bessie Smith, who regularly played in Chicago and throughout the East and the South. Chicago was a place where migrants from the rural areas of the southern states brought country blues and folk traditions with them including the blues players and performers. Carbine notes that blues and its performance tradition involved a direct interaction between performers and audience. Peyton's comment and the fact that live acts played jazz and blues on the same programme in the cinemas as the films screened demonstrate that those recently arrived migrants from the South were targeted and were part of the attraction of the house.

Chicago's African American vibrant musical and performance community provided a specific 'black cultural presence' that offers a historical example of a productive contrast between the regulating discourse of 'standardization' and the need for local performers to connect with their specific audiences. Our choice of a small theatre in Texas was somewhat more challenging. Jim Crow laws, the threat of legal and non-legal violence towards African Americans in the South would seem to mitigate against finding a similar set of cultural coordinates. One of the reasons for using a southern locale, though, was that the music we play, based around jug band and country blues songs, originated there. However, the history of a song we have played in the past called 'Deep Ellum Blues' provided a key to finding a historical example of just such a 'small theatre in a Texas town'. The song was originally recorded by The Lone Star Cowboys on the Victor and Bluebird labels in 1933. This version belongs to the kind of white proto-country style of the time and the treatment is up-tempo with guitar and mandolin and close harmonies. The lyrics of the song are specifically about this mythical space:

> When you go down in Deep Elem
> To have a little fun
> You better have your fifteen dollars
> When that policeman comes
> Oh sweet Mama, Daddy's got them Deep Elem blues

The song was claimed by another white band The Shelton Brothers from East Texas who recorded it for Decca in 1935. Both versions borrow, or actually steal, from an earlier song 'The Georgia Black Bottom' recorded by The Georgia Crackers for Okeh in 1927. However, the song is said to have been written by Blind Lemon Jefferson and Huddy Ledbetter (Leadbelly) around 1912 when they were together in the Deep Ellum area of Dallas Texas.[10] It tells of the temptations of areas where fun and sin are to be had and the pitfalls that await the pilgrim who seeks them out. When the white bands sing it, the reference is to a white man going to the black areas of town in search of 'illicit pleasures'.[11] 'Deep Ellum Blues' has since been recorded and played by any number of artists and bands from the original recordings, through the 1960s to the present day (we played Jerry Lee Lewis's jumped-up version). Once unhooked from its origins the song took on a life of its own, with added lyrics to the extent that it ceased to refer to any geographic space and occupied the libidinal Neverland of the popular, primarily white, conception of the blues.

As mythical as the song became, Deep Ellum is an actual area of Dallas, Texas that throughout the first half of the twentieth century was an African American shopping and business district and a well-known place for entertainment of all types, licit and illicit, throughout the outlying areas. The name comes from a derivation of Elm Street 'ellum', which may have come from either the East European Jewish immigrants or African American migrants who settled or passed through there.[12] It was located on a section of Elm Street in the South Dallas area at the Houston and Texas Central Railroad (H&TC) depot around which merchants and entrepreneurs had built businesses since the 1870s. Described by *The Chicago Defender* as 'the Broadway of the Dallas Black Belt' its reputation was widely known throughout the South. *The Defender* went on to add that the connecting Central Avenue area known as the Central Track was the 'Coney Island of the Negro District [...] where most of their amusements are staged'.[13] In the 1920s and 1930s it was a carnival of music and entertainment. 'Entertainment was presented outdoors and in streets and parks, dance halls, theatres and cafes as well as in "chock houses" and "soft drink stands" where bootleg alcohol was served.'[14] In that sense it bears similarity to areas like Storyville in New Orleans and Beale Street in Memphis as a hotbed of music and performance. Here all kinds of styles from W. C. Handy's band in the 1910s to the later urban jazz of Duke Ellington, the 'low down dirty blues' of Ida Cox and Bessie Smith in the 1920s, the 'hokum' of The Dallas String Band as well as street music of all kinds came together. It was a vibrant music culture, and at times a place where both black and white musicians might come together and play.

In the 1920s the same acts that played in Chicago, like Ellington and Cox played Deep Ellum's vaudeville theatres and dance halls. Two of these theatres, The Palace, manager Fred Hilson, and The Grand Central run by John Harris also ran film programmes. The Grand Central had a practice of playing films on the weekdays and bringing live acts and music in on the weekends. Both of these theatres advertised in the black newspaper *The Dallas Express* in the early 1920s. Nearby was the high-end black vaudeville Park Theatre run by Ella B. Moore. None of the programmes for the Grand Central I have seen list the orchestra members or give a sense of the type of instrumentation or musical selections for the films (Figure 6.1). Nevertheless the surrounding music and performance culture here suggests that the music played to films was indicative of the kind of styles that the audiences enjoyed throughout Deep Ellum.

A clearer record of what music cinema orchestras in Deep Ellum may have played can be found on the programme of Ella B. Moore's Park Theatre. Moore's Theatre was part of the black vaudeville circuit that existed in parallel to white vaudeville. In the Saturday, 4 September 1920 edition of *The Dallas Express* Moore advertised her programme as featuring 'Willie Toosweet and Sam Russell' and '12 All-Star People' (Figure 6.2). Willie Toosweet was a comic performer who throughout the 1910s had worked as a double act with 'Lulu' Toosweet; they were sometimes billed as 'Long Willie and Little Lula'. They were composers

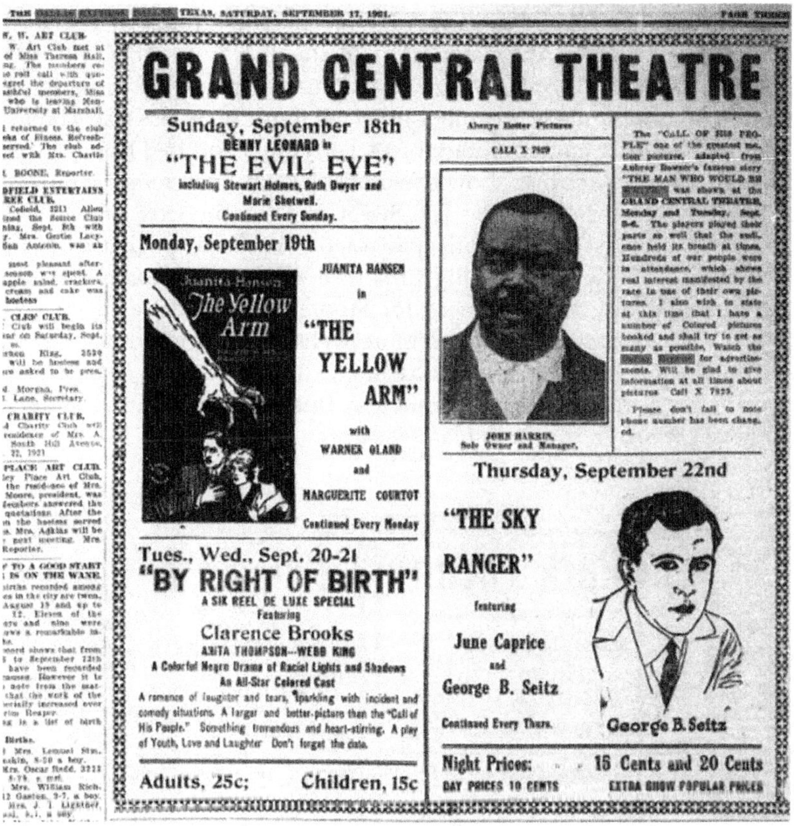

Figure 6.1 Advertisement for the Grand Central cinema in *The Dallas Express*, Saturday, 17 September 1921, p. 3.

as well as entertainers and were contemporaries of, and competitors with W. C. Handy. Lynn Abbott and Doug Seroff describe their act:

> The Too Sweets were a highly accomplished, multi-talented team. As performers they specialized in comic portrayals of children [...] Lulu was the prototypical 'baby soubrette' all childhood innocence one moment and then shamelessly manipulating her audience with a double-entendre blues [...] Willie played 'the booby', permitting all sorts of pranks to be played on him.[15]

On the same bill at The Park was Jimmie Cox 'The Black Chaplin', playing alongside Annie Mae Cox, who possibly played a similar role of soubrette that Edna Purviance did in Chaplin's films of the late teens; although given the description of the Too Sweets this was no doubt made relevant to the Deep Ellum audience through their kind of double-entendre comedy and blues-inflected music. Cox was also a blues composer, another contemporary of W. C. Handy, and had been a part of the Too Sweets' touring show based at the Gem Theatre in Memphis since the 1910s.[16] Both Willie Too Sweet and Jimmie Cox toured the South and the East in this early part of the 1910s through the early black-owned Vaudeville booking firm Fred Barasso's Tri State Circuit. By the time of the advertisement for Moore's Park Theatre they were most likely working through the more well-known and white-owned Theatre Owners Booking Association or T.O.B.A. which was responsible for bringing northern urban acts such as Duke Ellington, Ethel Waters and Bessie Smith to southern black theatres.

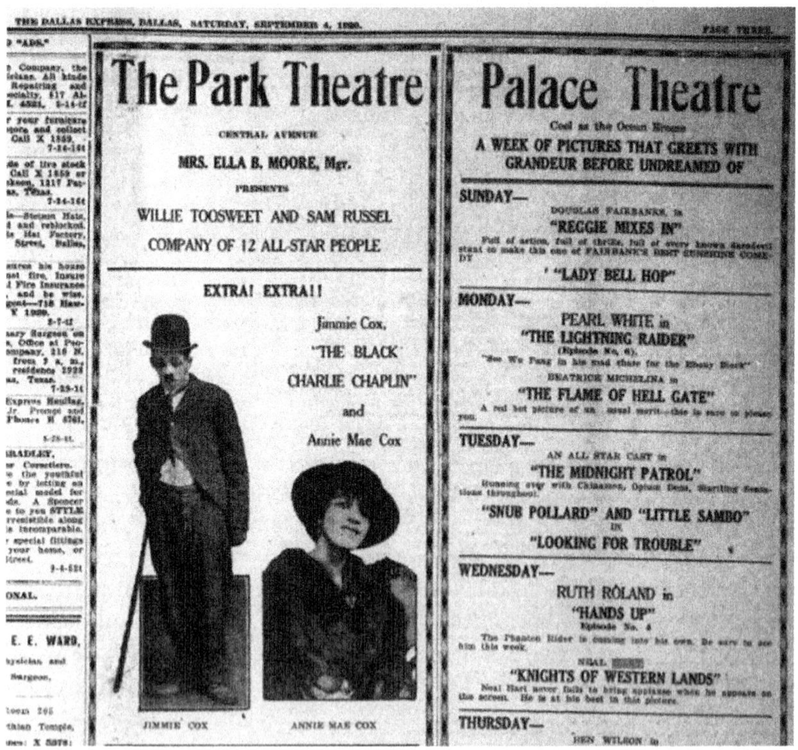

Figure 6.2 Advertisements for Ella B. Moore's The Park Theatre and the nearby cinema, The Palace in *The Dallas Express*, Saturday, 4 September 1920, p. 3.

Such a range of blues and jazz available to musicians and audiences in Deep Ellum stretched beyond a simple regional 'country blues' and outlines a picture of the significant role that black vaudeville played in shaping those forms. The role that popular theatre played in the development of blues forms has been suppressed in favour of a largely mythical image of the blues as a rural folk form. 'If Beale Street could really talk, it would speak not of a single father figure [W. C. Handy] but of a groundswell movement in which commercial possibilities for the blues were explored by a host of aspiring entertainers.'[17] In other words the music that surrounded The Palace and The Grand Central movie theatres in Deep Ellum was itself a combination of rural and urban blues, jazz and hokum styles that existed in a communal dialogue that criss-crossed each other, and this musical and performance 'landscape' gives some indication of the *probable* choices the musicians made to accompany the films.

We make it up as we go along

For me it is hard to imagine a more enticing or rich environment for film accompaniment of that time than that available to The Palace and The Grand Central, and that is precisely the kind of experience that guides our approach to accompanying the silent films we play. The picture of Deep Ellum on a Saturday night with medicine shows and string bands on the street and orchestras in the vaudeville theatres and cinemas is as intoxicating an image to me as it was to be there at the time... well maybe. In any case when we say 'a small cinema in a Texas town' we mean to render this carnival atmosphere for the film sound in the imagination of an audience in London ninety-five years later. Part of our showmanship is that we often announce that we make it up as we go along. This is somewhat disingenuous though in that while we do 'improvise', we do so through keeping an eye on cue sheets that I have devised and at the same time, and on a less conscious level, we draw on the traditions that were born out of environments like Deep Ellum and Memphis.

When Neil asked me to think about playing to the film with the band I had no real understanding of how most films in the silent period were accompanied. I was aware that cue sheets had been the norm in the teens and into the 1920s so I thought the best way to do this was to try to adapt songs and themes that we already knew to the scenes. So I devised my own cue sheet based on the intertitles (Figure 6.3). I worked from a grid of columns and rows. In the left cell is listed the intertitle. This is

Beggars of Life

Scene/Title	**Song and Key**	**Band Instrument**	**Piano**	
AS HE GESTURES HER TO COME WITH HIM	Vamp on Kassioe Jones, G	Dobro Tuned Guitar Washboard and Bass	**Joins in**	
AS THEY SIT DOWN TO LUNCH		FADE OUT	**Piano takes over**	FADE OUT
"Well, you gotta eat. I don't want you sick on my hands."	xxxxxx	xxxxxx		
"Me and my kid brother are tryin' to get home to our sick mother."	xxxxx	xxxxxx		
NEW SCENE: THEY'RE WALKING DOWN ROAD	**Slide guitar hints at poor Boy G**	XXXXX	**Piano follows**	

Figure 6.3 Cue sheet for William Wellman's *Beggars of Life* (1927).

then followed by the key and the theme in the next cell and then the next two cells are instrumentation with the final fifth cell to the far right for notes on treatment. The intertitles are there to keep us anchored in case we lose our place in the film.

White Oak is a Hart Western melodrama in which he plays, as he often did, a good 'bad' guy. As riverboat gambler Oak Miller he seeks revenge on the crook Mark Granger for deceiving his now ill sister Rose (Helen Holly) into giving up her virtue. Barbara (Vola Vale), Miller's love interest, has been nursing Rose who dies after telling Oak what has happened. Returning in disguise, Granger tries to abduct Barbara but is foiled by Oak, only to plot with Long Knife (Chief Standing Bear) to abduct her again while she is travelling west with her brother. A battle ensues with the Indians encircling the wagon train. A faithful dog rushes to tell Miller of the battle and he comes to the rescue. Long Knife is captured, and on discovering that Granger has also deceived his daughter he kills him.

Our instrumentation is along the lines that might have been found in The Grand Central, that is: guitars, mandolin, banjo, washboard, snare drum, double bass, harmonica and piano. There are five of us: Neil on piano, Mark plays the double bass and harmonica, Aly Hirji plays the mandolin and guitar, Alex Hammond plays the washboard and drums and does sound effects, I play the banjo, the dobro and the guitar and sometimes we sing. With our first effort, *White Oak*, we drew upon popular folk songs of the day such as 'Fair and Tender Ladies' for the love interest, 'Jack of Diamonds' for the gambling scenes and a rather raucous version of 'Brady and Duncan' for the battle scenes. Our style is built around the 'slap' bass where the note hits on the down beat, or the 'one' and 'three', and the 'click' hits on the upbeat, that is the 'two' and 'four'. While we do play waltzes, our rhythmic power lies in this 4/4 time. Our overall strategy was to break into these rhythms in the chase scenes, set a foreboding mood in the gambling scenes with the E minor key of 'Jack of Diamonds' and give over the intimate scenes between Oak and Barbara to Neil on the piano.

This first performance went over well with the London audience and was praised by the *Times Literary Supplement* reviewer as authentic. 'They are neither from Dodge city nor are they brothers but play like they are both.' Afterwards it was clear to me that we were products of the long history of standardization in our intent to serve the narrative of the film. Yet in some ways we unconsciously re-enacted some of the process of playing that had been done before the transition to the more standardized score sheet, in that we tried to hit changes in scenes and mood exactly. Neil's guiding principle was to keep the flow of the music going across the scenes and we combined applying specific songs to scenes with regard to mood alongside synchronizing with the action by 'effects'. That worked very well with the gunshots in the battle sequences for example, although an early scene where Oak jumps off the riverboat and swims was more difficult to judge. Altman's account of the late nickelodeon period points to a transitional moment around 1910 when the recommendations of film music columnists such as Clarence Sinn began to stress the continuity of music across scenes over the 'sound effects' approach. In this sense Altman says that the term 'cue' shifted from pointing to effects to recommending mood, therefore assuring continuity.[18] After that first performance, and through Neil's direction, we developed more confidence and were able to build the scene through improvising with dynamics and letting the musical phrase play out across scene changes, and thereby were more in line with the standardized form of supporting the story.

As much as I am a fan of W. S. Hart's *White Oak* it is not the most accessible of his films that I have seen. It suffers somewhat in its convoluted plot and pace, which for modern audiences creaks along. The next film we undertook, *Beggars of Life* directed by William Wellman in 1927, moves along at a more 'modern' rhythm. It also benefits from strong performances by Wallace Beery, Louise Brooks and David Arlen. Adapted from the 1924 Jim Tully novel *Beggars of Life: A Hobo Autobiography* it relays the real life of 'hobohemia' on the rails in 1920s America. Brooks plays Nancy, or 'the Girl', adopted from an orphanage, who shoots her guardian as he tries to molest her. The film opens just after the killing as 'the Boy' 'Jim' (David Arlen) knocks on their door. Convincing her she will be convicted of murder if she stays, she disguises herself as a boy and they run off to ride the rails. They come upon a hobo camp where a group of tramps are mixing up a 'mulligan stew'. To one side is a sick hobo being cared for by Black Mose, played by Edgar Washington. At the camp they meet Oklahoma Red (Beery) who is a hobo 'king'. 'The Girl' is discovered to be a woman by another hobo, The Arkansas Snake, which provokes a fight that is then interrupted by police who want to arrest her for the murder of her guardian. Red protects her, for his own nefarious purposes, and they run for a passing train. In a boxcar there is a mock trial over who is going to take possession of her. 'The Boy', Jim, tries to defend her but she saves herself by goading Red into fighting The Arkansas Snake whom he throws out of the boxcar door onto the tracks. The police board the train. The Boy and The Girl escape when Red uncouples the cars they are in from the main train and brings them to a stop just on the edge of a cliff. They make it to a cabin along with Black Mose and the sick hobo who later dies. Oklahoma Red finds them and seeing that they are truly in love he devises a scheme with Mose to dress the corpse of the hobo in Nancy's clothes and flat cap, put it on the railroad car, set it alight and run it off the cliff, thereby convincing the police that The Girl has died in the crash. When the police return to retrieve the cars Black Mose distracts them from the train as Red positions the corpse. Once the train is moving Red uncouples it again and sets it on fire and it gains momentum back towards the cliff. As he does this, Red is shot by the police. The Boy and The Girl escape first in Red's stolen Ford and then on top of another train on their way to the Canadian border. Wounded, Red falls off the train and dies on the cliffside in the last shot of the film.

The film was shot in the spring of 1928. The interiors were filmed at Paramount's studios in Los Angeles while the exteriors were shot on the railroad line that runs through Corrizo Canyon near Jacumba,

California, about seventy miles east of San Diego. It was released in two versions, a synchronized score and a silent version in order to serve the increasing number of larger film theatres that were equipping for sound. The synchronized score for the sound version was arranged by Emmanuel Baer and recorded between 20 August and 10 September 1928 at the Victor studios in Camden, New Jersey. Released on 22 September it was one of six Paramount releases that year that had a recorded soundtrack.

Baer had worked with Hugo Riesenfeld, the musical director for three Broadway theatres in New York City: The Rivoli, The Rialto, and The Criterion. In 1928 Baer had become freelance and moved towards composing for synchronized sound tracks. Riesenfeld was an arranger conductor whose techniques at the large theatres on Broadway, which at times had fifty-piece orchestras, was a blueprint for the standardization of film music in the 1920s. In her article 'The Presentation of Silent Films: Music as Anaesthesia' Gillian Anderson quotes Riesenfeld on his approach to arranging the musical accompaniment: 'I resort mainly for themes to the songs of the particular period of the film for which I am arranging the score [...] I usually divide the film into sections according to moods.' He had a room that held filing cabinets filled with music catalogued under the type of mood they evoked: 'sentimental, pastoral, heavy, dramatic [...]'. His process then was to go through each and find the appropriate piece for the scene. 'My course is to go through these and eliminate. My leading themes come in between these miscellaneous and may be as elaborate or as simple as the action requires.'[19] Baer, trained in this method, followed this same broad approach with *Beggars of Life*.

The film featured a specially composed song 'Beggars of Life', words by J. Keirn Brennan and music by Karl Hajos, and a song ('Hark those Bells') purportedly sung by Beery. The Victor paper records for the recordings listed in the Discography of American Historical Recordings (DAHR), however, show that there were four vocalists; William Cleary, who sang the title song, Donald Wells who sang in reel 4, which is the Beery song, with A. Ray and R. Moody as backup singers.[20] The files also show that they recorded music with orchestra to the films' nine reels plus a Prologue and an Epilogue. The Prologue features a recitation by Harrison Brockbank. Prologues were an accepted part of the Riesenfeld's film programme litany as well as that of most of the large prestigious urban theatres across the country. Both Prologue and Epilogue included the Brennan/Hajos song with the Prologue to be played prior to the screening of the film. This was accompanied by the Brockbank recitation, a scene-setting reading of hobo life, probably drawn from Jim Tully's prose. The Epilogue played out the film. We can assume that Baer followed the Reisenfeld method of employing the leading theme in various forms throughout the film.

The repetition of the main theme had a commercial value beyond the price of the cinema ticket. A trawl through the trade press of 1928 shows the impact the transition to sound, and the increasing popularity of radio and buying the records, was having on the industry. An article in the 26 September 1928 issue of Variety, entitled 'Mechanicals Overwhelm Music Business – 40 Theme Songs', noted that recorded music – 'the by-product of yesteryear' – was now to be the 'chief source of revenue' for the industry and that included 'sound pictures, records and radio'. It noted that forty theme songs had appeared in the previous two months, 'all hooked up with feature films and practically attaining mechanical recording with a minimum of investment by the publishers'.[21] Paramount was clearly on the bandwagon producing two songs, the title and the comic song 'Hark Those Bells' with Beery featured singing. These were heavily advertised in the film's promotion and theatres equipped with sound screened this version. To my knowledge *Beggars of Life*'s soundtrack has not been found and is probably lost. However, Harold 'Scrappy' Lambert's Brunswick recording of the title song does exist. Lambert, who later went on to sing with Red Nichols and his Five Pennies, recorded and released *Beggars of Life* in 1928. His version is in the sickly sweet 'crooner' style made popular by Rudy Vallee, with an orchestration of a small string section and piano. Compared to the original singer for the film, William Cleary, who in other recordings had a distinctive 'old timey' style, complete with dobro guitar, the Lambert version is clearly aiming for a wider market.[22]

While the larger white theatres in Dallas had geared up for sound by the time *Beggars of Life* was released, the smaller theatres such as The Grand Central or The Palace in Deep Ellum were yet to do so. This was not only true of black theatres, however. *Film Daily* ran a short item from their Dallas reporter: 'Of course the first run houses will reap the benefit at the beginning. But the little fellows will come in for their share when delivery and installation of equipment can be made more quickly and when the cost of these installations is more within their reach.'[23] While the article seems to be pointing to the inevitability of the sound film the studios were less sure. Paramount's release of two versions was designed to meet the shifting market, where demand for sound films had not fully taken hold. We play to the 16 mm print from the Eastman archive which has the song lyrics for 'Hark Those Bells' printed on title cards which probably means that we are playing to the silent version; although audiences might have been encouraged to sing along, which was a standard practice on the film programme. Like the environment for White Oak, the silent version would have been subject to a wider interpretation than that offered by the sound version.

In keeping with our method of accompaniment, which is to 'consult' history rather than try to reconstruct it, our approach to this film was to draw on the railroad songs that were popular throughout the early twentieth century. Given the atmosphere at either The Palace or The Grand Central in Deep Ellum many of these types of songs would have been circulating, either among the street performers, or by this time in the blues and jazz recordings that record companies like Brunswick, Vocalian and others had released. The songs we chose for this film were both blues and country folk based. Early in the film Jim watches Nancy try to hop a freight train. The train moves slowly from the water tower and as it gains momentum she runs after it but fails to reach it and falls to the ground. The motif we use for this comes from Walter E. 'Furry' Lewis's 1929 two-side recording of 'Kassee Jones' he did for Vocalion (subsidiary of Brunswick) in 1929. Lewis's version of this popular ballad seems to have been handed down through the railroad workers since its origins in the early 1900s and as Norm Cohen has demonstrated, it is a blues ballad and the words he uses have a number of sources in African American oral traditions such as 'Wreck of the Six Wheeler' and 'Hobo John'.[24] His loping rhythmic strumming provides an appropriate imitation of the driving wheels of the train as they speed up. Lewis had played with the Memphis Jug Band who did busk in the streets of Memphis; he played with W. C. Handy and was a part of the musical culture that was connected with black vaudeville throughout the South. Jug and string bands often engaged in the kind of 'hokum' of imitating farm animals and machinery and we follow suit in using the slow increasing rhythm of the washboard and muted strings on the acoustic guitar approximate the driving wheels while high string stabs on the lead guitar punctuate the whistle. Once the train picks up speed the double bass and piano come in.

This rhythm of the train underpins the film throughout but we open the credits with the Brennan/Hojas song 'Beggars of Life' (Figure 6.4). I play this on a single guitar and then the piano and double bass layer in. The verse chorus and refrain fits almost perfectly through the title sequence and the piano takes the fore across a fade to a shot

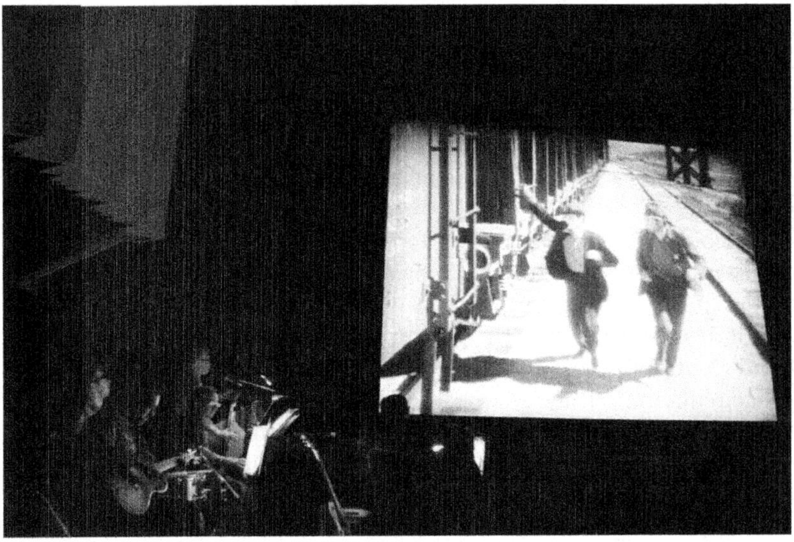

Figure 6.4 Neil Brand and the Dodge Brothers playing to William Wellman's *Beggars of Life* (1927), BFI Southbank, 10 April 2011.

of Jim's feet as he walks up to the screen door of Nancy's house and peers in. The sequence which follows is the discovery of the body and Nancy's retelling of the killing. This is a tense scene and we give it a sparse, effects-driven treatment building up to the train rhythm that will provide one of the main themes throughout the film.

This introduction allows us to foreshadow the rhythmic drive of the film and offers a contrast to the tender moments between Nancy and Jim. Tully's novel is almost devoid of any sentimentality and the romance plot was built into the film by Benjamin Glazer. Tully is recognized now as a forgotten founder of the hard-boiled American school of writing and his novel reveals a violent world of transient life on the road.[25] Written as an autobiography it is an early example of the road genre where characters meet on stops and byways provide colourful episodes but the motivation is simply to keep moving. The novel in this form did not fit into the classical Hollywood model easily and Glazer's 'couple-on-the-lam' solution allows the audience to sympathize with the plight of Nancy and the drifter Jim from the beginning. The hardboiled veneer is maintained throughout the first scenes until the third reel where they spend the night together inside a haystack. Here, sheltered from the cold and safe from the law, they confess to each other their loneliness and feelings of isolation:

> Jim: 'Some begs for one thing and some for another – and me, I ain't yet found out what I want.'
> Nancy: 'I know what I want – just a quiet place to be in – a place to keep clean in – a place to call home – I never had it... maybe I never will... I guess that's why I want it so.'

The dialogue, written by Glazer, functions to bring the couple together, delineates her desire to have a peaceful home which in turn prompts his role as protector but all the while acts as a thinly veiled suppression of his desire and love for her. The scene evokes a general trope of pathos in hobo and tramp narratives where the romance of the road is counterbalanced by its travails and hardships. Chaplin's incorporation of pathos into his comedy was only the most high-profile example in film culture at the time. The 'old timey' treatment of the title theme such as that of William Cleary as well as the updated crooning of Scrappy Lambert drip with sentiment. Other hobo songs from around that period such as Goeble Reeve's 'Hobo's Lullabye' or Jimmie Rodgers's 'Waitin' for a Train' are tinged with sentiment and wistfulness and would be adequate choices for this scene. The choice I made for this section came from another source, 'Po'Boy' by Booker T. Washington White, which is not strictly within the time frame as the recording we use was made by John and Alan Lomax at the State Penitentiary in Parchman, Mississippi (Parchman Farm) on 23 May 1939. Versions of this song, however, had been circulating since Sam Butler's (aka Bo Weavil Jackson) 'Poor Boy Blues' released in 1926 on Vocalion records. The White version is not a standard twelve-bar blues. He plays it on a dobro with a slide and with a rhythm that evokes the swaying movement of freight trains at a moderate speed. The melody is clear and haunting while the lyrics of the first two stanzas fit the mood of the scene:

> Po'boy. He was travelin' an' staying on the road
> Po'boy, he got hungry, he didn' even have nowhere to go

We slowed it down to match the contemplation of the dialogue and I wrote a bridge and melody which takes the motif one other place that works with the scene. Another liberty we take with this is that I sing a stanza right at the end in order to bring out the relationship and punctuate the scene. The lyrics I altered to:

> Poor boy's been runnin', ridin' blinds alone
> Poor boy's been lonesome and a long long way from home
> Poor girl's been fightin', trying to get on her own
> Poor girl's been fightin', to get a long long way from home

The scene is the emotional centre of the film and operates between the parameters of a hard-boiled hobohemia exteriority and a longing and vulnerable interior state of being. The melody of 'Po'Boy' evokes the sentiment of loneliness while the blues treatment maintains the masquerade of scepticism that the scene is designed to break through.

Constructing 'authenticity'

Neil Brand's idea of bringing a jug/skiffle band to play to these films has been an unexpected and enjoyable adventure. Our performances have been popular and widely praised for bringing an 'authentic' feel to the films, but perhaps more importantly they have brought the art of playing to silent cinema to younger audiences. The chance to write about this experience and to consider the history of the development of silent film music has also been illuminating. The invocation of a 'small theatre in Texas' began from a general understanding of the variety of often unruly environments that accompanied live music to silent cinema in the 1910s and 1920s. But on reflection, seeing these films while we play to them is nearer a modern twenty-first century experience. The audiences watch and, we hope, become immersed in the film while we try, as film composers from Riesenfeld to John Williams have, to keep the audience's attention on the narrative. We try to disappear. We do not attempt to approximate the exuberant effort that musicians playing to films on Chicago's south side or in Deep Ellum may have exerted. We can only speculate what African American players may have done with a film like Hart's *White Oak*, where the term 'white' in the advertising explicitly referred to his race. We do know that Hart films were a regular feature at The Grand Central, as were serials and mainstream Hollywood features, all replete with demeaning stereotypes of African Americans. We also know that films from black film companies such as the Lincoln Motion Picture Company, Micheaux Picture ['Film'] Corporation and the Norman Film Manufacturing Company were eagerly anticipated in the black press such as *The Chicago Defender, The Indianapolis Freeman* and *The Dallas Express*.

Some challenges exist in constructing cue sheets to films where representations of race, class and gender are retrograde. While we draw on performance and vernacular music traditions it is important that we draw the audience's attention to these aspects of the films when we introduce them. In *Beggars of Life* the character of Black Mose is played by Edgar Washington who plays his scenes with characteristic blackface minstrelsy mannerisms.[26] Towards the end of the film he plucks a chicken he has found while foraging and later he performs the minstrelsy trope 'frightened negro' in order to draw the police away from Oklahoma Red as he gets ready to set alight the lumber car containing the corpse dressed in Nancy's clothes. In both of these scenes we use the music to underpin the larger mood of the scenes. In the case of the chicken plucking, he does so without knowing that moments before in the same room Jim and Nancy have just watched his friend die. The scene is one of sadness and we play it that way. The ruse that he plays with the police is part of the final climactic scene and we use that to build up the tension rather than play minstrel melodies.

We participate in constructing an 'authenticity' which has little if anything to do with historical accuracy.[27] The songs we draw from, however, do have a relevance to the period. It is clear from Mary Carbine's work and from what I have learned of the entertainment culture of Deep Ellum, that the music played in cinemas was more likely to have reflected the tastes of the local audiences than abiding by the advice of a Clarence Sinn or playing from a more classical repertoire such as that of Riesenfeld and Baer. We also know that in Deep Ellum at least the audiences were not solely segregated.[28] If nothing else, this exercise helps to contribute to a more complex and complete history of film music in the silent period and hopefully to breathe life into the films for modern audiences.

Notes

1. Greil Marcus, *Invisible Republic: Bob Dylan's Basement Tapes* (London: Picador, 1997), 89, 94. Marcus is referring to the Harry Smith 1952 compilation of recordings *Anthology of American Folk Music*. In searching for a description of the musicians that appear here he notes that they were shut out of the dominant 'national narrative' and appear 'now like visitors from another world, like passengers on a ship that had drifted into the sea of the unwritten'.
2. Rick Altman, *Silent Film Sound* (New York: Columbia University Press, 2004), 390.

3. Ibid.
4. Ibid., 34; see also David Mayer, *Stagestruck Filmmaker: D. W. Griffith and the American Theatre* (Iowa City: Iowa University Press, 2009), 39–41.
5. Clarence Sinn, 'Music for the Picture', *Motion Picture World*, vol. 26 (November 1910), 1227, quoted in Altman, *Silent Film Sound*, 243.
6. Harold B. Franklin, *Motion Picture Theatre Management* (New York: George H. Doran Company, 1927), 282.
7. Mary Carbine, 'The Finest Outside the Loop: Motion Picture Exhibition in Chicago's Black Metropolis, 1905–1928', *Camera Obscura*, vol. 23 (May 1990), reprinted in *Silent Film*, ed. Richard Abel (New Brunswick: Rutgers University Press, 1996), 234–62 (251).
8. Carbine, 'The Finest Outside the Loop', 252.
9. Ibid., 254.
10. Kevin Pask, 'Deep Elum Blues', *Southern Spaces*, 30 October 2007. http://southernspaces.org/2007/deep-ellum-blues#section3, accessed 26 February 2015.
11. Alan B. Govenar and Jay F. Brakefield, *Deep Ellum: The Other Side of Dallas* (College Station: Texas A&M University Press, 2013), 20–1.
12. Ibid., 5.
13. Ibid., 6.
14. Ibid., 3.
15. Lynn Abbot and Doug Seroff, '"They Cert'ly Sound Good to Me": Sheet Music, Southern Vaudeville and the Commercial Ascendency of the Blues', in *Ramblin' On My Mind: New Perspectives on the Blues*, ed. David Evans (Urbana: University of Illinois Press, 2008), 49–104 (75). Quotes within are from 'The Profession at Jacksonville, Fla', *The (Indianapolis) Freeman*, 14 May 1910, 7.
16. Abbot and Seroff, '"They Cert'ly Sound Good to Me"', 75.
17. Ibid., 81.
18. Altman, *Silent Film Sound*, 220.
19. Gillian Anderson, 'The Presentation of Silent Films: Or, Music as Anaesthesia', *Journal of Musicology*, vol. 5, no. 2 (Spring, 1987), 257–95 (270).
20. Emmanuel Baer entry, Discography of American Historical Recordings. http://adp.library.ucsb.edu/index.php/talent/detail/17666/Baer_Emmanuel_director, accessed 10 March 2004.
21. 'Mechanicals Overwhelm Music Business – 40 Theme Songs', *Variety*, 26 September 1929, 56.
22. William Cleary who recorded for Perfect records in a duet with Arthur Cornwall 'Carolina's Lyric Tenor' in 1930. There are three of their recordings that exist: 'The Old Rugged Cross', 'When They Ring Those Golden Bells' and 'Where We Never Grow Old', for Perfect Records, all sentimental religious songs aimed at a rural market, gospel tunes played in an 'old timey' style.
23. 'They Say – "Sound in Small Towns"', from *Motion Picture Times*, Dallas, reprinted in *Film Daily*, Sunday, 19 August 1928, 5.
24. Norm Cohen, *Long Steel Rail: The Railroad in American Folksong*, 2nd edition (Urbana: University of Illinois Press, 2000), 146.
25. Charles Willeford, 'Jim Tully: Holistic Barbarian', Introduction to *Beggars of Life: A Hobo Autobiography* (Edinburgh/London/Oakland: AK Press/Nabat, 2004, orig. pub. New York: Albert & Charles Boni, 1924), i–viii.

26. Edgar 'Blue' Washington was a star in the African American baseball leagues in the 1910s before he became a film actor. His career spanned the silent and sound period from Harold Lloyd shorts such as *Haunted Spooks* in 1919 to an uncredited role in *The Hustler* in 1960. He appeared with John Wayne in *Haunted Gold* (1932) but was often uncredited in the seventy films that IMDB list; other reports state that he was in up to 100 films. See Mark V. Perkins, 'Edgar 'Blue' Washington', Society for American Baseball Biography Project. http://sabr.org/bioproj/person/b347deac, accessed 3 March 2015.
27. 'Authenticity' in this context derives from a set of cultural assumptions established in the middle of the twentieth century by what Paige A. McGinley has described as a 'cult of authenticity' in which 'cultural brokers – critics, record producers [...] labored to establish American blues music and musicians as authentic, and therefore, valuable'. Paige A. McGinley, *Singing the Blues: From Tent Shows to Tourism* (Durham, NC and London: Duke University Press, 2014), 7.
28. Govenar and Brakefield note that there were midnight shows for white people at Ella Moore's Park Theatre, where admissions were higher than those paid by black audiences (*Deep Ellum*, 60–1).

Bibliography

Abbot, Lynn and Doug Seroff, ' "They Cert'ly Sound Good to Me": Sheet Music, Southern Vaudeville and the Commercial Ascendency of the Blues', in *Ramblin' On My Mind: New Perspectives on the Blues*, ed. David Evans (Urbana IL: University of Illinois Press, 2008).

Abel, Richard (ed.), *Silent Film* (New Brunswick: Rutgers University Press, 1996).

Altman, Rick, *Silent Film Sound* (New York: Columbia University Press, 2004).

Anderson, Gillian B., 'The Presentation of Silent Films: Or, Music as Anaesthesia', *Journal of Musicology*, vol. 5, no. 2 (Spring 1987).

Carbine, Mary, 'The Finest Outside the Loop: Motion Picture Exhibition in Chicago's Black Metropolis, 1905–1928', *Camera Obscura*, vol. 23 (May 1990), 9–41.

Cohen, Norm, *Long Steel Rail: The Railroad in American Folksong*, 2nd edn. (Urbana, IL: University of Illinois Press, 2000).

Franklin, Harold B., *Motion Picture Theatre Management* (New York: George H. Doran Company, 1927).

Govenar, Alan B. and Jay F. Brakefield, *Deep Ellum: The Other Side of Dallas* (College Station, TX: Texas A&M University Press, 2013).

Marcus, Greil, *Invisible Republic: Bob Dylan's Basement Tapes* (London: Picador, 1997).

Mayer, David, *Stagestruck Filmmaker: D. W. Griffith and the American Theatre* (Iowa City: Iowa University Press, 2009).

McGinley, Paige A., *Singing the Blues: From Tent Shows to Tourism* (Durham, NC and London: Duke University Press, 2014).

'Mechanicals Overwhelm Music Business – 40 Theme Songs', *Variety*, 26 September 1929.

Pask, Kevin, 'Deep Elum Blues', *Southern Spaces*, 30 October 2007. http://southernspaces.org/2007/deep-ellum-blues#section3.

Perkins, Mark V., 'Edgar "Blue" Washington', *Society for American Baseball Biography Project.* http://sabr.org/bioproj/person/b347deac.
Sinn, Clarence, 'Music for the Picture', *Motion Picture World*, vol. 26 (November 1910).
'They Say – "Sound in Small Towns"', *Film Daily*, Sunday, 19 August 1928.
Willeford, Charles, 'Jim Tully: Holistic Barbarian', Introduction to *Beggars of Life: A Hobo Autobiography* (Edinburgh/London/Oakland: AK Press/Nabat, 2004) (orig. pub. New York: Albert & Charles Boni, 1924).

Part II

Novel Music and New Issues

7
Bringing a Little Munich Disco to Babelsberg: Giorgio Moroder's Score for *Metropolis*

Jeff Smith

From the 1910s to the late 1920s, producers and distributors sought several means of standardizing musical accompaniments for silent films.[1] Trade press columns, cue sheets and original scores all provided guidance to the pianists and orchestras charged with supplying dramatically appropriate music to films. The circumstances of exhibition, however, encouraged a range of different responses to these efforts to unify silent film accompaniment practices. For example, some accompanists lacked the talent or technical facility required to play a composed score or perform the music suggested by a cue sheet. In other cases, exhibitors rejected the music supplied by producers and distributors because they preferred to cater to audience tastes or to promote locally popular performers. In still other situations, economic incentives dictated the exhibitor's musical choices as he or she featured pieces 'plugged' by a publisher or retailer.

The mere fact that exhibitors were free to reject music specially prepared for particular titles has had long-term consequences for the ways they are seen in modern contexts. Indeed, during the silent era, individual accompanists or conductors could choose to highlight one aspect of a scene over another. One might elect to highlight the setting, another could underline specific character traits, and still another might play the scene's overall mood. These options endowed silent films with a certain semiotic flexibility when it comes to music, and this, in turn, has enabled modern composers and performers to graft their unique style onto the vision of the original filmmakers. When the musician is someone like Philip Glass, who appeals to a cultural elite, the resulting collaboration can seem like an intermedial experiment that showcases the strengths of both artists. When, however, the musician comes from a more popular or mainstream background, the resulting collaboration can feel like a form of hubris where the artist's desire to enhance or improve the effect of a canonized masterpiece feels faintly ridiculous. The latter proved true in the case of Giorgio Moroder's score for the 1984 re-release of Fritz Lang's *Metropolis* (1927). As Liam Lacey wrote in a review of *Metropolis*'s soundtrack album, 'With its spare arrangements and wide-screen production, this is all-too-typical Moroder fare [...] What's next? *Birth of A Nation* to Journey's greatest hits?'[2] A. O. Scott was even blunter in his assessment of Moroder's version of *Metropolis,* calling it a 'desecration' nearly twenty years after it was released.[3]

Moroder's warrant for replacing Gottfried Huppertz's original score for *Metropolis* was the notion that the film's futuristic setting could motivate more modern musical sounds. A key component to Moroder's musical palette was the synthesizer, an instrument he already had featured in both his pop recordings and his film work. Indeed, when compared to the traditional rock band's line-up of guitar, bass, piano and drums, the synthesizer's icy electronic textures much more readily fit the machine aesthetic that was central to *Metropolis*'s imagery.

Yet Moroder's synth-pop sound also informs his assembly of the soundtrack at a much deeper level, serving as an overarching metaphor for his project as a whole. By combining old images with new music and traditional scoring techniques with more modern ones, Moroder's *Metropolis* aims for a synthesis of various strands of film and musical culture. In bringing Lang's classic film into the present, Moroder partially bridged the gap between the time of its production and that of its monumentally constructed world. More than that, Moroder's music itself sits at the crossroads of several strands of early 1980s pop, including disco, Album Oriented Rock (AOR), and the New Romantics of the British New Wave. The musical hybridity evident in Moroder's soundtrack for *Metropolis* reflexively underlines the fact that musical accompaniments for silent films were an inherently heterogeneous mixture of different forms and styles. In what follows, I provide both a brief production history of Moroder's version of *Metropolis* and an analysis of the dramatic functions of his score. Despite the mostly tepid critical response to his work, Moroder provided a template that many later musicians would follow in their own attempts to modernize silent cinema.

Crafting the UFA-pop sound: Moroder remounts *Metropolis*

During the disco era, Giorgio Moroder established himself as one of the industry's pre-eminent producers of dance music. Moroder's career began in the German discothèque scene in the early 1970s. He played guitar and bass with several bands in the Munich area, but his big breakthrough came in 1974 when he joined with songwriter Pete Bellotte and singer Donna Summer to produce the latter's first album, *Lady of the Night*. Moroder and Bellotte followed up by producing a string of disco classics with Summer, including 'Love to Love You Baby', 'Bad Girls', 'Hot Stuff' and 'Dim All the Lights'. Perhaps the most important record produced by this team was *I Feel Love*, a proto-synth-pop icon that featured a pulsing Moog bassline and a sequencer-derived rhythm track. Even today, the song feels years ahead of its time, essentially creating the template that many later techno musicians would follow in the late 1980s and early 1990s.

Although Moroder's work with Summer made him famous within music circles, he also branched out by making solo records and doing film work. In 1978, Moroder wrote the Oscar-winning score for Alan Parker's prison picture, *Midnight Express*. The award increased the demand for his services both as a film composer and songwriter. Working with singer Debbie Harry, Moroder co-wrote 'Call Me' for Paul Schrader's *American Gigolo* (1980). The track became one of Blondie's biggest hits, reaching number one on *Billboard*'s singles charts. Next Moroder composed both the score and title song for Schrader's *Cat People* (1982), this time collaborating with singer David Bowie. (The song made a notable reappearance in Quentin Tarantino's *Inglourious Basterds* (2009) underscoring the scene where Shoshanna prepares to turn her cinema into a death-trap for the Nazi high command.) In 1984, Moroder won his second Oscar, along with Keith Forsey and Irene Cara, as the co-writer of 'Flashdance... What a Feeling'. The record topped *Billboard*'s singles charts and sold more than a million copies. Even more impressively, the soundtrack from *Flashdance* went on to sell more than five million copies within its first year in release. Fuelled by Moroder's music, the film itself proved to be a great success for its distributor, Paramount, grossing more than $200 million worldwide on a budget of only $4 million.

Moroder's success as a record producer and film composer earned him the opportunity to develop the kind of vanity project often associated with major Hollywood stars. In Moroder's case, he set his sights on *Metropolis*, Fritz Lang's portrait of a dystopian future where economic inequalities have caused most of Metropolis's citizenry to toil in an inhumane work environment below the city's surface. A 'one percenter' narrative years before its time, *Metropolis* depicts a society where the labour of the many enables the wealth and leisure of the few. Freder, the son of Joh Fredersen, a powerful industrialist, enjoys a life of privilege and hedonistic pleasure in gardens and arenas located on Metropolis's rooftops. Freder's interest in Maria, a daughter of one of the workers, leads him into the city's depths where he becomes aware of the workers' exploitation. Freder later attends a meeting led by Maria, who retells the parable of the Tower of Babel as a means of providing the workers psychological succour. As R. L. Rutsky points out, within *Metropolis*'s vaguely Christian symbolism, Maria 'personifies the emotional and spiritual aspects of life that have been excluded from Fredersen's rationalized, functional world'.[4]

Meanwhile, Fredersen conspires with a mad scientist named Rotwang to discredit Maria by creating a robotic *doppelgänger*. After kidnapping the real Maria, Rotwang sends his robot into Metropolis in an effort to undermine the former's spiritual authority. Roused by the false Maria, the workers run roughshod in the city's lower levels, demolishing the various machines that oppress them. The resulting destruction, though, causes the underground areas to flood, endangering all of the workers' children. The workers eventually turn on the false Maria, burning her at the stake as though she were a witch. In the film's final scenes, Freder chases Rotwang to the city's rooftops. After a furious battle, the mad scientist falls to his death. In the final shots, Maria encourages Freder to act as an intermediary between Joh and the workers, serving as the symbolic 'heart' that unifies Metropolis's 'head' and 'hands'.

According to a *New York Times* report, Moroder's interest in *Metropolis* was spurred by an executive at Paramount, who suggested that he 'do something with a silent movie'.[5] Further impetus was supplied by the then recent success of Francis Coppola's reconstruction of Abel Gance's *Napoleon* (1927).[6] Moroder eventually screened more than twenty different titles in search of the right project.[7] He apparently had long been a fan of *Metropolis*, but wanted to be sure that there wasn't a better film out there, just waiting for the Moroder touch.

Once he settled on *Metropolis*, Moroder then made a fifteen-minute test film to see whether his idea of adding new music to Lang's film was workable.[8] Moroder always believed that the film didn't really need a classical score, and thus was confident that *Metropolis*'s narrative and style would support the addition of newly composed pop music. During a recording session with David Bowie, though, Moroder was surprised to learn that the singer had similar ambitions of reviving *Metropolis*. The two collaborators ended up competing with one another for the film rights. Moroder eventually won out, paying a tidy sum of $200,000.[9]

Within six months, Moroder's idea of simply adding new music to an existing silent film suddenly got more complicated. He learned from

Enno Patalas of Munich's Film Museum that the Library of Canberra had some missing footage from *Metropolis*. Moroder's negotiation with the Australians to use approximately eight new scenes led to further discoveries. A film collector in Los Angeles possessed a nitrate print of *Metropolis* while the Censorship Ministry in Berlin enabled Moroder to consult the film's original cue cards and music score, the latter containing the conductor's handwritten notations.[10] By now, the nature of Moroder's project had changed; the composer not only intended to rescore the film, but restore it as well.

Although the material uncovered by Moroder enabled him to construct what was then the most complete version of Lang's film, still another discovery gave rise to the most controversial editorial decision the composer would make. Following a tip about a collector in San Diego, Moroder examined a 9.5 mm copy of *Metropolis* that used subtitles for the film's dialogue rather than printed title cards. Moroder decided to replicate this strategy in his own reconstruction, claiming that the film simply flowed better without the dialogue intertitles.[11] Clearly this was a significant deviation from Lang's original, but it seems likely that Moroder opted to use subtitles as a means of making the film more accessible to the young viewers he'd hoped to entice into theatres.

During the two years Moroder spent tracking down missing elements from *Metropolis*, he also began work on the score. Initially, he prepared a purely instrumental accompaniment to the film and previewed it to a small group. Said Moroder, 'They had a hard time understanding it.' Going back to the drawing board, Moroder then added four new songs to the music score he previously had composed. In developing these songs, Moroder supplied the music while the singers, including Adam Ant, Yes's Jon Anderson, and Queen's Freddie Mercury, provided lyrics. He then showed this new version to a test audience of 400 people. Apparently the songs improved the viewers' uptake of the story. But Moroder also wanted to know whether the film ought to have more or fewer songs. Eighty per cent said the film should have more. After seeing such a clearly expressed preference for additional songs, Moroder hooked up with his old lyricist, Pete Bellotte, and the pair contributed four more tunes to the film's soundtrack.[12] This last batch also featured some of the era's biggest recording stars, such as Pat Benatar and Bonnie Tyler.

With the soundtrack completed, Moroder readied his new version of *Metropolis* for its world première at the Cannes Film Festival. Included as a sidebar, the film debuted on 15 May 1984. As Todd McCarthy wrote in *Variety*, Moroder's new version of *Metropolis* received a mixed response from the Cannes crowd with buffs complaining about the composer's throbbing, contemporary score and 'more pragmatic industryites' saying that the modern music might help a new generation of viewers connect with the film.[13] Although many gave Moroder credit for compiling the most complete version of *Metropolis* currently available, one festivalgoer quipped, 'that if Fritz Lang isn't rolling in his grave, then the late Henri Langlois, who outlawed any musical scores for silent films at the Cinematheque Francais, most certainly is'.[14]

Meanwhile, shortly after the Cannes première, Moroder was already gearing up for a follow-up film project. *Variety* columnist Army Archerd reported in July of 1984 that Moroder had signed a deal with MGM/UA to write and direct a new musical. Moroder envisioned the film as an opportunity to showcase new, unknown talent. He also promised that the music, sound, dialogue and effects would all be digital.[15]

A little more than a month after its appearance at Cannes, Producers Sales Organization, who handled worldwide sales of Moroder's *Metropolis*, announced that Cinecom International had acquired the US rights. The film's American première, a benefit performance of *Metropolis* for the Astoria Film Restoration Society, was already planned for early August. A theatrical release would follow with the film opening in twelve key cities, gradually expanding to another thirty venues about three weeks later. Producers Sales Organization's president, Mark Damon, explained the slow rollout: 'Cinecom saw the film not as an art house picture, but as one that could attract a youth audience because of its score and become a cult classic.'[16] Damon added that Moroder's new version already had been sold in 80 per cent of its foreign territories with openings planned in Japan, Italy, Belgium and Australia.[17]

A little more than a week after the American Museum of the Moving Image benefit in Astoria, *Metropolis* screened at yet another high-profile benefit, this time at the Academy of Motion Picture Arts & Sciences' Samuel Goldwyn Theater. With the proceeds going to the Academy Foundation's education and preservation programme, the show also featured a short documentary about the film's reconstruction that was produced by Moroder and directed by Academy archivist Daniel Woodruff. According to *Variety*, the presentation of *Metropolis* was especially notable since it marked the first time a feature film was shown with a digital soundtrack on a digital stereo sound system.[18] Dave Concors, *Metropolis*'s technical director, worked with Dolby engineers to install the new system at the Goldwyn. It utilized an SMPTE time code from an interlock magnetic reproducer to synchronize the image with a Sony digital 24-track reproducer for playback.[19] Touting the quality of this system's sound reproduction, Moroder said, 'This will sound so good that it will make people think about improving the sound of films. The irony is that it took a silent movie to have the first digitally reproduced score.'[20]

The day after the Academy show, Producers Sales Organization took out an ad in *Variety* congratulating Moroder, Cinecom, and French distributor Coline/Gaumont for *Metropolis*'s successful New York and Paris openings. The ad featured blurbs from popular newsweeklies, like *Newsday*, and infotainment programmes, like *Entertainment Tonight*. It also reported a five-day gross of $34,368 at New York's 57th Street Playhouse and a seven-day total of $103,560 at five cinemas in Paris.[21] These numbers were impressive considering the film's limited release. However, those kinds of box office returns would have to continue if Moroder was to recoup his $2 million investment.

By November of 1984, though, *Metropolis*'s re-release had only earned back about $635,000 in domestic release. According to Cinecom, *Metropolis* did slightly better abroad, earning just under a million dollars in foreign box office.[22] But any way you sliced it, Moroder's project was a financial flop. By March of 1985, Cinecom was readying the film for the non-theatrical market, attempting to soak up what few bits and bobs of revenue remained.

Cinecom, though, had not given up on the core audience that it believed would embrace *Metropolis* as a cult film. On 1 March 1985, the Campus Network beamed *Metropolis* to four colleges in the first satellite to theatre transmission of a feature film. The campuses involved in this screening included the State University of New York–Fredonia, the University of North Dakota–Grand Forks, and the University of Wisconsin–Oshkosh. One of the chief attractions of this type of delivery system for *Metropolis* was the fact that the satellite transmitted Moroder's digitally recorded, 24-channel stereo soundtrack, which was then played back at each venue in four-channel stereo.[23]

With more complete versions of *Metropolis* currently available, Moroder's restoration of Lang's film looks more and more like a historical curiosity. Moroder's music track may have seemed cutting edge at the time, but the mix of disco, 'hair metal', and Adam Ant now suffuses the film with an 'I Love the Eighties' vibe, endowing the final product with the airless quality of a time capsule. Since it is no longer defensible on preservationist grounds, the long-term legacy of Moroder's *Metropolis* might well rest with its technological innovations in digital sound playback and satellite transmission.

An analysis of Moroder's score

The film begins with a brief prologue. A pair of superimposed titles explain the setting and basic premise of *Metropolis* followed by additional supertitles that provide background information about Moroder's restoration. Even prior to the appearance of these titles, though, the first musical track in Moroder's score, 'Machines', gradually sneaks into the soundtrack under a black screen. The song is a mid-tempo synth-pop track featuring a pulsing rhythm. When the first supertitle appears, the texture of the music thickens with the addition of a simple, syncopated melody. Although an instrumental, 'Machines' evokes other British bands of the era such as Human League. Indeed, Moroder's composition would not sound out of place on *Dare*, their commercial smash released just a couple of years earlier.

When *Metropolis*'s title card appears, though, the rhythmic drive of the music halts, eventually giving way to a series of rapid arpeggios. This essentially serves as a brief musical bridge into the film's opening scene. These arpeggios fade out and are replaced by fast, repetitive synthesized percussion. Moroder uses this opening as an opportunity to create an aural collage of musical effects that loosely links up to the montage of moving machine parts that follows. In a subtle homage to Lang's early sound period, these repetitive rhythms evoke the noises made by heavy machinery in the famous opening scene of *The Testament of Dr. Mabuse* (1933). When Lang cuts to an insert of a clock on the wall, Moroder adds a musical 'tick-tock' sound that explicitly mimics the movement of its hands. On a cut to a long shot of several whistles blowing, Moroder inserts a dissonant chord as a sort of musical 'stinger'. That sound carries over into the next shot showing dozens of workers preparing for a shift change.

As one crowd of labourers gets off an elevator and another gets on, Cycle V's 'Blood From a Stone' sneaks onto the soundtrack. With its chanted rising melody, the song captures the dour mood of the workers as they move lockstep together, functioning as human cogs in a larger industrial infrastructure. The lyrics further reinforce the dehumanized condition of this workforce, doomed to toil underground. Phrases like 'human chain' and 'every hour like the last' conjure up the drudgery that is the labourer's lot in *Metropolis*. Moreover, the reference to mercy's hidden face reinforces the overall impression that the workforce is an impersonal, anonymous mass. 'Blood From a Stone' eventually segues into the sound of white noise and walla, anticipating the scenic shift to the rooftop stadium where we are first introduced to Freder. As a son of the chosen few, Freder is shown starting a race, participating in physical culture that serves as an explicit contrast to the manual labourers' blood and sweat.

This series of music cues covers only the first five minutes of *Metropolis,* but nonetheless serves as a guide to the score's larger musical structure. Moroder's music essentially oscillates between songs and aural collages of synthesized sound effects, relying on brief snatches of instrumental music to knit these disparate elements together. The collages primarily have a depictive quality, mimicking bits of visual action and selectively cueing certain diegetic sounds heard by the characters onscreen as part of the fictional space. In an interview with *Billboard,* Moroder differentiated his work on *Metropolis* with his previous film scores, saying, 'I'm not only putting the music to a movie, but also sound effects and words.'[24]

Here Moroder avails himself of the synthesizer's ability to function as a sort of noise machine. For example, to accompany a series of cityscapes showing forms of modern transportation, Moroder layers together synthesized sounds to suggest various kinds of urban noise pollution, such as car horns, helicopter rotors and train engine motors. At other times, whoosh sounds are synched with picture to match the blasts of steam emitted by large machines. Later, in the scene where Joh Fredersen first sees Rotwang's robot, a vaguely metallic sound matches each of the android's steps as she walks towards Joh. And when the Grim Reaper swings his scythe during Freder's hallucination, Moroder's synthesizer plays pitch-bending portamentos that synch up with the blade's every swipe through the air.

By precisely synching musical effects to visual actions, Moroder engages in techniques akin to the kind of 'mickey-mousing' made famous by Max Steiner and other classical Hollywood composers. Moroder's willingness to incorporate synthesized noises, however, also underlines the essentially hybrid character of *Metropolis*'s soundtrack. These scenes recall nothing so much as the synchronized music scores and 'part-talkies' that appeared during the transition between the silent and sound eras.

Moroder's songs, on the other hand, play some of the same sorts of functions seen in other pop music scores of the early 1980s. They signal shifts in mood, reinforce *Metropolis*'s larger thematic ideas and give voice to feelings that the characters may not explicitly verbalize. For example, when Maria enters Freder's pleasure garden, Pat Benatar's 'Here's My Heart' foreshadows her place as part of the 'head/hands/heart' allegory that analogizes the city's class structure and spatial divisions to the stratified parts of a human body. At a more immediate level, the music also explicates Freder's sexual attraction to Maria. Although Freder merely asks who she is, Moroder's song hints at his unspoken feeling, thereby providing motivation for his search for her.

Moroder uses 'Here's My Heart' in a loosely leitmotific fashion, reprising the song as a theme for Maria in scenes showing the development of her relationship with Freder. The song returns when Freder approaches Maria after she recounts the legend of Babel to a rapt audience of workers. It also is reprised at the very end, covering the final scene where Freder serves as an intermediary between Joh and the workers.

In a somewhat similar fashion, Bonnie Tyler's 'Here She Comes' serves as a musical theme for the robot Maria created by Rotwang. The song is introduced when Freder first encounters the false Maria outside his laboratory. It then comes back a few minutes later to accompany the scene where the false Maria foments rebellion among the workers.

Moroder's voice casting of the two singers reinforces the difference in traits exhibited by Maria and her false double. Although Pat Benatar established her reputation as throaty, hard-rock shouter on records like 'Heartbreaker' and 'Hit Me With Your Best Shot', her number in *Metropolis* – 'Here's My Heart' – shows off her softer side, exploiting the coloratura parts of her vocal range. By utilizing an implicitly feminine singing style, Benatar's performance matches the gentle concern and sincerity that Maria exudes as a spokesperson for the workers and their families.[25] In contrast, Bonnie Tyler became known for her husky, vocal rasp on records like 'It's a Heartache' and 'Total Eclipse of the Heart'. Although some of those qualities are slightly obscured by the submergence of her voice in Moroder's mix, its grain is still evident in the scenes where 'Here She Comes' underscores the robot's actions. Tyler's craggy voice thus highlights the false Maria's more carnal and sensual dimensions, traits showcased more fully during her celebratory erotic dance at the Yoshiwara Club. As the only two female singers that appear on *Metropolis*'s soundtrack, Moroder neatly matches the semiotic qualities of Pat Benatar's and Bonnie Tyler's vocal personalities to the specific attributes of the two Marias.

The leitmotific functions of Moroder's songs are especially evident in Benatar's and Tyler's numbers, but other pieces are treated in a similar fashion. For instance, Moroder's 'Machines', which first appears in *Metropolis*'s prologue, returns later when the false Maria leads a worker uprising. The music thus underscores the mob's attack on the objects that serve as metonymic representations of the Metropolis's larger structures of oppression. 'Machines' is played a third time when the massive workers' revolt collides with the crowd of revellers departing from the Yoshiwara Club. It continues under the scene where the workers prepare to burn Rotwang's robot, meting out vengeance on the last machine still functioning in Metropolis. Indeed, the music briefly stops as the fire reveals the metallic endoskeleton beneath the false Maria's skin. But it picks up again to underscore the climactic chase where Freder rescues the real Maria from Rotwang's clutches. Here Moroder uses the song to capture the pace and intensity of Freder's pursuit and his rooftop battle with Rotwang. It ends abruptly, though, when the latter stumbles off a balcony, and 'Machines' is replaced by a descending pitch that 'mickey-mouses' Rotwang's fall.

Jon Anderson's 'Cage of Freedom' also serves as something of a leitmotif for Joh Fredersen, 'the Master of Metropolis'. The song is introduced in the scene where Freder expresses concern about his father's treatment of the 'hands' who built the city. The song's title and lyrics loosely convey the paradox that characterizes Joh's situation. Although he essentially rules over Metropolis, his elevated spatial and social position has left him 'imprisoned' in a de facto cell of his own design. The song returns later when Joh sees the city's electricity go out and realizes that the system he built is on the verge of collapse. Joh's assistant reminds him that Freder is in the city's depths with the rebelling workers. Joh's concern for his son's safety ultimately draws him out of his secure lair, forcing him to finally make contact with the workers he has exploited.

Although most of Moroder's songs for *Metropolis* are primarily motivated by character, a few others are motivated by situation. Loverboy's 'Destruction', for example, furnishes rather 'on the nose' accompaniment to the scenes of the workers attacking their foreman Grot and rising up against the machines. It continues as the real Maria quickly responds to the floodwaters beginning to rise within the city's depths. Similarly, Freddie Mercury's 'Love Kills' underscores a brief scene where two men, driven by sexual attraction to the false Maria, scuffle over the prospect of earning her affections. This is followed by elliptical cuts to two other men fencing and another shooting himself in the head. Like 'Destruction', 'Love Kills' is an example of the kind of 'lyrical literalism' that engendered criticism in scholarly discussions of early MTV videos.[26] Yet the use of 'Love Kills' also imposes a kind of unity on these disconnected episodes, indicating the desire and despair felt by those who fail to gain possession of the false Maria's body.

Perhaps Moroder's most unusual move in his score for *Metropolis* was the decision to use only sound effects for the scene where Maria and Freder lead the workers' children away from the rising floodtides.

No music is heard during this scene. Instead we hear only rushing water, crowd walla, and the beat of the gong-like object that alerts Freder and another worker to the danger. Moroder's choice was unconventional insofar as the scene contains moments that typically motivate underscore, such as Freder and Maria's brief reconciliation and their last-minute rescue of the children. Indeed, it is hard to imagine a theatre organist in the 1920s laying off such obvious opportunities to play up the scene's drama.

Accompaniment guides and scores, however, indicate that there were earlier precedents for such uses of musical silence. Edith Lang and George West's *Musical Accompaniment of Moving Pictures: A Practical Guide for Pianists and Organists* allowed for the use of brief musical silences as a sort of pause for dramatic effect, a technique they claim was virtually mandated by scenes of death.[27] Moreover, Eric Dienstfrey's research on Louis F. Gottschalk's score for *The Patchwork Girl of Oz* indicates that some scenes were shown without any type of musical accompaniment.[28]

These earlier instances of musical silence further point up the degree to which Moroder's score replicates the essentially hybrid character of silent film accompaniment. Although Moroder's musical idiom is quite different, his use of songs, bridges, aural collages and silence works as a correlate to the earlier period's techniques. Indeed, one could easily imagine a pianist at the time of *Metropolis*'s initial release predating Moroder's strategy by accompanying the film with songs selected from the Hit Parade of 1927. In this hypothetical scenario, Paul Whiteman and Sophie Tucker would take the place of Freddie Mercury and Pat Benatar. Yet despite obvious changes in musical styles, the underlying logic guiding such choices remains largely the same.

Conclusion

When Moroder's *Metropolis* was finally released in 1984, even those who defended it did so on grounds that had little to do with his unusual score. Some reviewers praised Moroder's restoration of Lang's film noting that it was more complete and looked better than previous versions. Others treated the music as a 'gateway drug', something to get the kids in the door to see *Metropolis* in the hope that they would eventually get off on Lang's trippy visuals. The release of more complete restorations of *Metropolis* in 2002 and 2010, however, almost immediately rendered Moroder's version anachronistic. Among the things 'restored' in these versions was Gottfried Huppertz's original score. Lacking its earlier imprimatur, Moroder's *Metropolis* now seems more and more like a historical curiosity.

Does this mean that Moroder's *Metropolis* is doomed to be swept into the dustbins of history, the slipstreams of our minds? I think not, largely because Moroder had the foresight to procure the rights to the film. This enabled him to create 35 mm prints of his version for theatrical distribution, which, in turn, allowed the film to have an afterlife in ancillary markets. The circulation of Moroder's *Metropolis* on videotape, DVD, and Blu-ray is something often denied more evanescent experiments in silent film accompaniment, and it has earned his version some vestige of cultural relevance.

Beyond its availability in these different distribution windows, Moroder's score for *Metropolis* remains influential for an additional reason. The Munich disco master demonstrated that silent film accompaniment could be a space for intensely personal expressions of musical style. Whether one sees Moroder's music as an example of overweening ambition or as an effective match for Lang's futuristic imagery, there is little question that it perfectly encapsulates the sound he had been developing for more than a decade. Using the synthesizer as his primary tool, Moroder fashioned an amalgam of disco, rock and German electronic music that was uniquely his own. By grafting his vision atop Lang's, Moroder encouraged countless others to explore new paths and possibilities for modernizing the sound of the 'silent' cinema.

Notes

1. For more on silent film accompaniment practices, see Rick Altman, *Silent Film Sound* (New York: Columbia University Press, 2007); Gillian B. Anderson, *Music for Silent Films, 1894–1929: A Guide* (Washington, DC: Library of Congress, 1988); Kathryn Kalinak, *Settling the Score: Music and the Classical Hollywood Film* (Madison: University of Wisconsin Press, 1994), 40–65; and Martin Miller Marks, *Music and the Silent Film: Contexts and Case Studies, 1895–1924* (New York: Oxford University Press, 1997).
2. Liam Lacey, 'INSIDE THE SLEEVE POP Metropolis: Original Soundtrack Giorgio Moroder/Various Artists', *The Globe and Mail*, 6 September 1984, E3.
3. A. O. Scott, 'A Restored German Classic of Futuristic Angst', *New York Times*, 12 July 2002, E17.
4. R. L. Rutsky, '*Metropolis*', in *Film Analysis: A Norton Reader*, ed. Jeffrey Geiger and R. L. Rutsky (New York: W. W. Norton, 2005), 185. For a slightly different take on Lang's film, see Tom Gunning, *The Films of Fritz Lang: Allegories of Vision and Modernity* (London: BFI, 2000), 52–83.
5. Annette Insdorf, 'A Silent Classic Gets Some 80's Music', *New York Times*, 5 August 1984, H15. See also Janet Maslin, 'Hey, Don't Forget the Audience', *New York Times*, 19 August 1984, H15.

6. Ibid.
7. Ibid.
8. Insdorf, 'A Silent Classic Gets Some 80's Music'.
9. Ibid.
10. Ibid.
11. Ibid.
12. Ibid.
13. Todd McCarthy, 'New "Metropolis" Sparks Controversy at Cannes', *Variety*, 16 May 1984, 4.
14. Ibid.
15. Army Archerd, 'Just for Variety', *Variety*, 26 July 1984, 3.
16. James Greenberg, 'Reconstructed "Metropolis" Release Set', *Variety*, 28 June 1984, 3.
17. Ibid.
18. 'Digital Sound for "Metropolis" at Pic Academy', *Variety*, 16 August 1984, 6. It is worth pointing out that Moroder had long been established as an early adopter of new recording technologies. He pioneered the use of digital recording on his solo album *E = MC2*. In an interview with *Rolling Stone*, however, he acknowledged that this experiment was something of a failure: 'My last album was an expensive mistake: I was so concerned with the technology of digital recording that I ignored the creative angle, and it really shows.' See Steve Pond, 'Giorgio Moroder is Looking for a New Sound', *Rolling Stone*, 30 October 1980, 28; and Jim McCullaugh, 'Digital for Disco Via Moroder LP', *Billboard*, 22 September 1979, 38.
19. Pond, 'Giorgio Moroder is Looking for a New Sound'.
20. Insdorf, 'A Silent Classic Gets Some 80's Music'.
21. See *Variety*, 17 August 1984, 9.
22. Jim Robbins, 'Cinecom Posts '84 Profit, Branches Into Production', *Variety*, 23 January 1985, 12.
23. 'Satellite-to-Theater Pic Delivery Service Bows on U.S. Campuses', *Variety*, 13 March 1985, 20.
24. ' "Good Old Disco" Still Works for Moroder', *Billboard*, 29 October 1983, 74.
25. Interestingly, in a 1980 interview, Moroder expressed a passing interest in working with Benatar. Describing the types of new recording stars that he'd like to groom, Moroder emphasized the importance of a singer's appearance, saying 'it is much easier if a girl not only can sing, but is good looking, like Pat Benatar or Blondie'. See Pond, 'Giorgio Moroder is Looking for a New Sound', 28.
26. See, for example, Andrew Goodwin's *Dancing in the Distraction Factory: Music Television and Popular Culture* (Minneapolis: University of Minnesota Press, 1992), especially 66–7 and 85–8.
27. Edith Lang and George West, *Musical Accompaniment of Moving Pictures* (Boston: Boston Music Company, 1920), 58.
28. Eric Dienstfrey, 'Synch Holes and Patchwork in Early Feature-Film Scores', *Music and the Moving Image*, vol. 7, no. 1 (Spring 2014), 40–53. These silent passages in *Oz*, though, were nowhere near as lengthy as the scene in *Metropolis*, which runs a little more than three and half minutes without music.

Bibliography

Altman, Rick, *Silent Film Sound* (New York: Columbia University Press, 2007).

Anderson, Gillian B., *Music for Silent Films, 1894–1929: A Guide* (Washington DC: Library of Congress, 1988).

Archerd, Army, 'Just for Variety', *Variety*, 26 July 1984.

Dienstfrey, Eric, 'Synch Holes and Patchwork in Early Feature Film Scores', *Music and the Moving Image*, vol. 7, no. 1 (Spring 2014), 40–53.

'Digital Sound for "Metropolis" at Pic Academy', *Variety*, 16 August 1984.

'"Good Old Disco" Still Works for Moroder', *Billboard*, 29 October 1983.

Goodwin, Andrew, *Dancing in the Distraction Factory: Music Television and Popular Culture* (Minneapolis: University of Minnesota Press, 1992).

Greenberg, James, 'Reconstructed "Metropolis" Release Set', *Variety*, 28 June 1984.

Gunning, Tom, *The Films of Fritz Lang: Allegories of Vision and Modernity* (London: BFI, 2000).

Insdorf, Annette, 'A Silent Classic Gets Some 80's Music', *New York Times*, 5 August 1984.

Kalinak, Kathryn, *Settling the Score: Music and the Classical Hollywood Film* (Madison, WI: University of Wisconsin Press, 1994).

Lacey, Liam, 'INSIDE THE SLEEVE POP Metropolis: Original Soundtrack Giorgio Moroder/Various Artists', *The Globe and Mail*, 6 September 1984.

Lang, Edith Lang and George West, *Musical Accompaniment of Moving Pictures* (Boston, MA: Boston Music Company, 1920).

Marks, Martin Miller, *Music and the Silent Film: Contexts and Case Studies, 1895–1924* (New York: Oxford University Press, 1997).

Maslin, Janet, 'Hey, Don't Forget the Audience', *New York Times*, 19 August 1984.

McCarthy, Todd, 'New "Metropolis" Sparks Controversy at Cannes', *Variety*, 16 May 1984.

McCullaugh, Jim, 'Digital for Disco Via Moroder LP', *Billboard*, 22 September 1979.

Pond, Steve, 'Giorgio Moroder is Looking for a New Sound', *Rolling Stone*, 30 October 1980.

Robbins, Jim, 'Cinecom Posts '84 Profit, Branches Into Production', *Variety*, 23 January 1985.

Rutsky, R. L., '*Metropolis*', in *Film Analysis: A Norton Reader*, ed. Jeffrey Geiger and R. L. Rutsky (NewYork: W. W. Norton, 2005).

'Satellite-to-Theater Pic Delivery Service Bows on U.S. Campuses', *Variety*, 13 March 1985.

Scott, A. O., 'A Restored German Classic of Futuristic Angst', *New York Times*, 12 July 2002.

Variety, 17 August 1984.

8
Soviet Fidelity and the Pet Shop Boys

Beth Carroll

'Only a contrapuntal use of sound in relation to the visual montage piece will afford a new potentiality of montage development and perfection'; so said Sergei Eisenstein, Vsevolod Pudovkin and Grigori Alexandrov in 'A Statement on Sound' (1928).[1] Writing at a time of vast developments in film sound they expressed a fear that 'sound recording [would] proceed on a naturalistic level, exactly corresponding with the movement on the screen, and providing a certain "illusion" of talking people, of audible objects, etc.'.[2] For these Soviet filmmakers and theorists, sound should shy away from synchronization and move towards a juxtaposition of the image and sound track in order to best serve montage. Eisenstein, Pudovkin and Alexandrov were assured in their belief that montage was still the most effective means of creating an emotionally and symbolically powerful film. As Peter Larsen describes:

> For Eisenstein [...] it was a question of retaining the experimental European montage culture that had functioned throughout the silent film period as an artistic alternative to mainstream films. For this reason they reject the synchronisation of sound and image, calling instead for 'a contrapuntal use of sound in relation to the visual montage piece'. They imagine that in time this can lead to 'an orchestral counterpoint of visual and aural images'.[3]

Tatiana Egorova details how such writings and theorizing on film music in the Soviet Union developed within the context of debates surrounding how film should be experienced and, indeed, who should be manipulating this experience. Regarding early Soviet film music accompaniment, Egorova argues it arose out of an understanding that music not only helped actors deliver more expressive performances, in terms of movement and gesture, but, following the same logic, could help audiences watch the film.[4] As Kracauer suggested, music enhances the impressive effect of silent sequences.[5] Interestingly, however, early Soviet film music accompaniment, marked by a dependence on pianists who mediated the filmic text and provided 'the technical and, partially, the aesthetic conditions for watching', moved towards original musical scores due to the ubiquity of orchestras.[6] 'The need to create original music scores for silent films', Egorova claims, 'became especially acute in the latter half of the 1920s; it was connected with the fact that a considerable number of cinemas in the large towns of the country had symphony orchestras at their disposal.'[7] Eisenstein, Pudovkin and Alexandrov's 'A Statement on Sound' (1928) thus can be understood as occurring at a time that marked an increasing number of films having originally composed musical soundtracks. The result of this is quite wide-ranging and, in part, leads to issues of authenticity, fidelity and 'originality'. For, following traditional views of authorship, once an original soundtrack has been composed either in consultation with or under the blessing of the director, surely we have a 'definitive' audio-visual experience endorsed by the filmmaker? Any subsequent reworking of the whole, or part, of the experience thus cannot escape the discourse of adaptation.

Through a focus on reviews of the Pet Shop Boys' soundtrack album to Sergei Eisenstein's *Battleship Potemkin* (1925) alongside reviews of their live performance which accompanied screenings of the film, this chapter will explore issues surrounding the notion of fidelity. It will examine whether there were any differences in reception between the album as a stand-alone creative work and when accompanied by the film and experienced as an audio-visual event. It will place the reviewers' discourse within the context of Eisenstein's own writings on sound, issues of adaptation and the understanding of the canon. An interview with the Pet Shop Boys will make clear the group's own feelings towards their soundtrack, their approach to its creation and the background to the project.[8]

Predominantly known for their focus upon electronic, synth-pop music, the Pet Shop Boys, consisting of Neil Tennant and Chris Lowe, formed in 1981 and have to date released twenty-three albums.[9] The band's use of electronic instruments coupled with their being influenced by the disco genre has fuelled a debate around the, rather nebulous, issue of musical 'authenticity'. Mark Butler, for example, has described the style of the Pet Shop Boys in the following terms:

> Lead singer Neil Tennant's vocal delivery is remarkably smooth; he avoids noisy elements such as exhalations, pronounces words crisply, and rarely varies the dynamic level. In fact, this lack of exertion has been a dominant theme in critiques of the Pet Shop Boys' music. [...] Such critiques almost always mention the Pet Shop Boys' use of synthesizers and drum machines as well.[10]

The perceived lack of 'effort' is here clearly linked to the idea of authenticity, and places the Pet Shop Boys' music which references camp and false urbanity, in direct contrast with the discourses surrounding rock music. The latter often focused on the image of a single performer that displayed great effort in order to convey authentic expression. As Butler explains, 'it was important that [the individual performer] be seen as the author of the sounds she created; displays of effort were a way of highlighting this relationship'.[11] This contrasts greatly with the use of musical technology employed in electronic music, and epitomized by the Pet Shop Boys; 'In electronic dance music, however, technology often obscures the relationship between music and its creator [...] In some cases, synthesizers and drum machines become the performers, and the creator of the sound only performs once, when he or she programmes the music.'[12]

Fred Maus is in accordance with these ideas but takes them further and links them with the notion of community:

> In folk revival discussions, the notion of authenticity encapsulated the belief that certain music is valuable because it grows out of, and embodies, values of a community; subsequent discussions of rock and other popular music retained the notion, altering it to include truthful self-expression as well as reflection of a community.[13]

As the interview with the Pet Shop Boys makes clear, the project behind the soundtrack and live performances started at the request of the Director of the Institute of Contemporary Arts. The finished product resulted in a varied musical palette that features the trademark sound of the electronic score and featuring the Dresden Symphony Orchestra. The use of an orchestra, whilst not new to the Pet Shop Boys' music, could be seen to reference back to that early musical period in Soviet cinema that Egorova describes. The Pet Shop Boys' music for *Battleship Potemkin* features a strong mix of compelling rhythmic beats, electronic and percussive, with melodies, strings, trumpets, sound effects and occasional lyrics; a rich sonic landscape. D. Travers Scott in his piece titled

'Intertextuality as "Resonance": Masculinity and Anticapitalism in Pet Shop Boys' Score for *Battleship Potemkin*', has provided a helpful breakdown of the soundtrack listings with its accompanying scenes.[14] He describes how the soundtrack 'continues juxtaposing and integrating these elements of classical, choral and dance genres. In so doing, their commonalities are highlighted; their sympathies are amplified, most in terms of sentiment rather than rhythm, style or instrumentation. The unity that coheres the soundtrack is one of feeling.'[15]

The critical reception of the Pet Shop Boys' album *Battleship Potemkin*, released in 2005 by Parlophone and EMI Classics, was chiefly positive. One review by *Drowned in Sound* described the album as effecting a strong emotional and sensorial response in the listener:

> It's impressive, weighty music with immense emotional power, winding the tension to a pitch in such a casual, natural manner that you don't realise you've been holding your breath until the sound crashes and you breathe out again. Or, alternatively, dropping the tension in favour of an emotion-besieging melancholic ache which welds the persistent beats of electro to the drawn-out ache of the string section and proceeds to hold the listeners [sic] emotions utterly hostage without any apparent effort.[16]

Of interest here to the reviewer is the music's ability to engage the listener on a physical, rather than an explicitly intellectual, level; the typical electronic style of the Pet Shop Boys being key to this compositional achievement. One could create a parallel here between the sensory engagement the reviewer felt with the music and Eisenstein's own views of montage.

Eisenstein, famous not only as a filmmaker but also a theoretician, developed the theory of montage whereby individual elements, when placed together, create new meaning; much like a collage.[17] David Bordwell highlights how '[I]n his theoretical writings [...] Eisenstein considers both verbal and visual metaphors to be extensions of basic kinesthetic ones.'[18] These metaphors had the aim of eliciting emotional reactions in the audience. As Bordwell notes, '*Potemkin* subordinates experimentation to the interests of emotional exhortation.'[19]

D. Travers Scott outlines the audio-visual relationship between the soundtrack and Eisenstein's film.[20] In his piece Scott argues that there is a parallel to be drawn between the revolutionary principles of Eisenstein's *Battleship Potemkin* and the alternative masculinity represented by the Pet Shop Boys' musical oeuvre. He places this debate within the larger framework of 'resonance' which he defines as the 'active, sympathetic amplification and clarification of commonalities'.[21] As Scott highlights, issues of resonance fit within the wider discourse on adaptation, more particularly intertextuality.

Adaptation, the debates around which appear predominantly in English Studies focusing largely upon a movement between literature and stage/screen, often occurs in terms of binaries. As Shelley Cobb states when reflecting upon this travel from novel to film: '[m]uch of the critical literature on adaptation continues to reflect expectations that films should simply translate their source material, and a film's value is assessed according to how faithfully it reproduces the original text'.[22] This canon buttressing approach, whilst having provided fecund discussion on adaptation, is ultimately reductive and offers limited opportunity for understanding adaptations: 'Not only does fidelity criticism limit the critic, it also limits the text, confining the adaptation to a compare-and-contrast analysis.'[23] Using Robert Stam's 'Beyond Fidelity: The Dialogics of Adaptation', Cobb states that '[a] dialogical approach to adaptations moves beyond the binary relationship of novel and film to recognizing the importance of context to intertextuality'.[24] More useful to both adaptation studies as a whole and our discussion on the Pet Shop Boys' soundtrack for *Battleship Potemkin* in particular, is a move away from traditional notions of fidelity towards the concepts of translation, intertextuality, the impacts of cultural contexts, and production, amongst other factors.

Existing as a stand-alone album as well as within the context of a live audio-visual event, the Pet Shop Boys' soundtrack to *Battleship Potemkin* sits within this spectrum of discussions surrounding adaptation. Robert Stam has argued that 'the source novel [...] can be seen as a situated utterance, produced in one medium and in one historical and social context, and later transformed into another, equally situated utterance, produced in a different context and relayed through a different medium'.[25] Whilst this is in many respects an apt description of the Pet Shop Boys' soundtrack, there are certain issues to highlight. Firstly, what is the Pet Shop Boys' soundtrack a transposition of? Secondly, is this an example of a transcoding across different conventions? Undeniably linked, these two questions illustrate the porous boundaries that surround adaptations.

The first issue to address, however, is the noticeable absence of such issues in *Drowned in Sound*'s review. Focusing rather upon the emotional response the music elicits (thereby complementing Eisenstein's own feeling towards film and montage), the review tends to eschew discussion of the background to the soundtrack and indeed offers little in the way of musical analysis. The quality of the review notwithstanding, it is interesting to note two things: firstly, that whilst the potentially audio-visual relationship between the soundtrack and Eisenstein's film is not of interest, the reviewer clearly enjoys the album and holds it in esteem, and secondly, in their analysis they highlight the Pet Shop Boys' industrial sound:

> The songs are largely instrumental, mixing the heavy, unnerving industrial threat often displayed by Kraftwerk with the soaring melancholy of strings provided by the Dresdener Sinfoniker. This makes for a sound both spacious and intense, the kind of sound which fills the whole room and monopolises the ear's attention in a manner which one really doesn't expect from a purpose-written soundtrack.[26]

The reviewer Sarah McDonnell has sentiments in a similar vein and states that, 'While it's hard to say from the soundtrack alone how well it compliments [sic] the film, the music is certainly strong enough to stand up without visual aid.'[27] For McDonnell the soundtrack has every right to exist as a stand-alone piece that fits within the wider musical framework of the Pet Shop Boys' oeuvre. She continues, '[m]any tracks see the traditionally lush Tennant/Lowe sound stripped right back to simple, electronic motifs. Yes there is an orchestra present, but the influence of electronica is far more prevalent throughout the soundtrack.'[28]

As I have already discussed the Pet Shop Boys' sonic style and raised the issue of 'authenticity' that such a style elicits, it is salient here to return to the issue of adaptation. One could argue that the 'industrial threat' *Drowned in Sound* highlights, which is likened to the group Kraftwerk, illustrates a parallel between the Pet Shop Boys' music more widely (and the soundtrack more specifically) and the thematic and aesthetic qualities of not only *Battleship Potemkin* but to Eisenstein's filmography more generally. Indeed when discussing the musical accompaniment to the original release of *Battleship Potemkin* Tatiana Egorova has described how:

> When [*Battleship Potemkin*] was on in Berlin, the German composer Edmund Meisel approached Eisenstein with an offer to write music for it. The artistic effect surpassed all expectations: Meisel's music formed such an organic unity with the visual sequence of the picture that Leon Feuchtwanger, shaken by what he saw,

> devoted a whole chapter of his novel *Erfolg* (*Success*) to a description of the effect produced by both the representation and the music in *Battleship Potemkin* upon the thousands of Berliners. Afterwards, when analysing his film, Eisenstein confessed that *Battleship Potemkin*, though shot as a silent film, 'had much of what can be achieved by a sound picture'. The effect achieved was greatest in the scene where the battleship encounters the whole squadron, and 'music of the machines' ruled supreme: 'In style, this film already overstepped the limits of "a silent picture illustrated by music" and passed into the new province of *tone-film*.'[29]

As Egorova claims, Eisenstein himself applauded Meisel's original soundtrack to *Battleship Potemkin* for its ability to unify the music with the mechanical visuals. Important to Eisenstein's admiration was undoubtedly the rhythmic harmony between image and sound; between the images of mechanical movement, industry and technology, and the music's ability to replicate this through a process of media transcoding. The Pet Shop Boys' music, whilst suffering accusations of inauthenticity, a moniker they are happy to engage with ironically, is here able to provide an apt parallel with Eisenstein's images; they provide a 'faithful' analogous musical counterpart to the image. As such, we can begin to see corresponding thematic clarity between the two texts.

Scott argues likewise and suggests that *Battleship Potemkin*'s focus upon the 1905 revolt of Russian sailors against authority can be likened to the Pet Shop Boys' 'cultural work around gender, in which they articulate alternatives to dominant constructions of masculinity'.[30] He, like Fred Maus and Mark Butler before him, explores the way in which the Pet Shop Boys engage with alternative culture. This engagement is in part due to a ubiquitous ambiguity inherent to their work. Maus, whilst discussing many of the fans' interpretations of such songs as 'Go West' (released in 1993) and 'It's a Sin' (1987) within the context of the Gay Debate ('whether gay listeners have special insight into the songs') states that 'Like much communication about homosexuality, Tennant's lyrics are often double-voiced, carrying special meaning for insiders while remaining differently meaningful for others as well.'[31] This ambiguity creates the possibility of an oppositional stance within the music, and it is through this opposition that further equivalences can be drawn. As Maus states: 'Generally, ambiguity or ambivalence involves setting out an opposition, along with an unclear or undecidable relation between the terms of the opposition, and such patterns are ubiquitous in the Pet Shop Boys' music.'[32] Thus if we return to the question of what the Pet Shop Boys' soundtrack is a transposition of, we can understand the answer as a palimpsest. The soundtrack on one level can be read as 'remake' of Edmund Meisel's 'original' soundtrack to the film, on a secondary level as an 'adaptation' of the visual images of *Battleship Potemkin,* and finally as a transcoding or translation of the thematic and tonal properties of Eisenstein's film. This certainly confuses the issue of fidelity which, whilst a contradictory subject within adaptation studies, is often still valorized in terms of 'faithfulness' to the 'original text'.

Interestingly, whilst such discourse is absent in the reviews of the Pet Shop Boys' soundtrack as a stand-alone album, it is ubiquitous in reviews of the live performance.[33] Maddy Costa, for instance, writing in *The Guardian* stated that

> [T]he Pet Shop Boys appeared so awed that the most they could provide was a portentous swirl of synthesised minor chords. At times the music felt positively inappropriate, particularly when a pulsing dance beat accompanied the captain's attack on his sailors, and Tennant began chanting the word 'Ya'.[34]

It would not be a stretch to infer that the awe that the Pet Shop Boys are supposed to have felt during the live performance is due in large part to some form of reverence towards Eisenstein's film. Whilst reverence towards *Battleship Potemkin,* implied or actual, may exist, the rhetoric that Costa is utilizing works towards reaffirming a canonical discourse whereby Eisenstein's film is venerated and admired as art, an 'original' which forms a point of comparison for the Pet Shop Boys' soundtrack with little regard for differences in artistic forms. Paul Bond, in his review titled 'Failing Eisenstein: The Pet Shop Boys' New Score for Battleship Potemkin in Trafalgar Square', whilst still reaffirming the place of *Battleship Potemkin,* does, however, appear to provide a more nuanced approach:

> There is nothing inherently wrong in the idea that a synthesiser pop duo – who have a certain reputation for intelligent and ironic lyrics and an avant garde sensibility – might be able to produce a contemporary score for a silent film. It would be quite possible to produce a score with such instruments, which have a certain flexi bility and range. The requirement, as with any instrumentation or style of composition in this context, must be that the music enhance

> and augment the film. The music should not dominate the film, nor should it betray the film's vision. It can rather, as Eisenstein seems to have suggested, renew a film's contemporary resonance.[35]

Bond, like many critics and theorists, has not only made the film the primary element in this audio-visual relationship but reaffirmed the position of sound and music being secondary to the image more generally. The result is that Bond seems conflicted as to what the soundtrack should strive to do or what the Pet Shop Boys' music achieves in this marriage. 'Sometimes they succeeded', he states, '[m]uch of the time the score was unobtrusive and unexceptional, and allowed the film to move at its own pace. Occasionally, a small melody rose out of the rhythm, and there was some general agreement between sound and vision.'[36] The soundtrack is thus caught between allowing the film to progress in an unobstructed manner and providing more overt musical moments.

This discourse on unobtrusively permitting the film to remain the dominant text reductively dismisses understanding of energy transferral and artistic transcoding into different conventions. The result of which can be seen in the following rhetoric:

> To produce a score for Battleship Potemkin, however, it is necessary to understand the film. This would involve an engagement with the political conceptions that drove Eisenstein to create this tribute to the heroism of 1905. It would involve an understanding of 1905, and of the lessons that were learnt from it. It would require an honest assessment of the explosion of revolutionary events after 1917. And, I would add, it must also involve a degree of understanding of the betrayal of 1917 by the Stalinist bureaucracy, which was able to assert its power only through the physical destruction of communists – so that one avoids the easy portrayal of the idealism that animated the Potemkin sailors and Eisenstein himself as simply utopian dreaming. What we got, though, was something else entirely. There was the possibility of bringing one of the greatest works of art of the twentieth century into a sharp new focus. The event, though, was geared towards the Pet Shop Boys, and their lack of understanding of the film militated against such a possibility from the start.[37]

Not only is Bond reaffirming the film's place in the canon of high art but he places this alongside the promotion of intellectual engagement as the only true form of understanding. The superciliousness of this position notwithstanding, it does present an irony. Namely that, if Bond had started from an intellectual comprehension of the Pet Shop Boys' music, both thematically and musically, he would have been able to draw parallels between the two works that would demonstrate the artistic exchange taking place through the process of adaptation, such as those described by Scott. However, akin to conservative debates surrounding fidelity more generally, rewarding discussion has been dismissed in favour of re-establishing a cultural and artistic hierarchy. Analogous to the considerations surrounding authenticity in the Pet Shop Boys' work, their wider stylistic devices and use of genres, reviewers of the live performance of *Battleship Potemkin* have failed to see the possibility of cultural exchange and commentary present within the work.

Through the use of electronic music and alternative masculinities, the Pet Shop Boys' soundtrack actively engages with the thematic and formal qualities of Eisenstein's *Battleship Potemkin* and, one might argue, in such a way that Eisenstein himself would compliment. Whilst exceedingly pleased with Meisel's soundtrack for the film, Eisenstein far from believed the film a finished product; rather he favoured the idea of the soundtrack being rewritten every ten years.[38] As a revolutionary film, it is the intertextual cultural references and juxtaposition of different artistic expressions that permit the film to remain as salient today as in 1925. The Pet Shop Boys' soundtrack's electronic score, which is at times melodious, at others pulsing, creates the very contrapuntal soundtrack that enables audiences to interpret the film in innovative ways.

Interview with the Pet Shop Boys

Q: Can you give some background to the project? How did you get involved? Did you have any misgivings?

A: We were approached by Philip Dodd, then director of the Institute of Contemporary Arts. He had been given the opportunity to stage an event in Trafalgar Square by the then Mayor of London, Ken Livingstone, and decided to screen *Battleship Potemkin* because the story of the film resonated with the long tradition of public political protest in the Square.

Q: Would *Battleship Potemkin* have been your choice of film to score?

A: It wasn't our decision.

Q: Prior to this project, had you done much work that required composing to pre-existing visuals?

A: No, none.

Q: What was your thought process in terms of how to approach such a project? Did you do much research into previous soundtrack re-imaginings? Did you read around the subject?

A: We watched the film on DVD (turning down the Shostakovich music that was dubbed on). Then we asked our programmer to load the DVD onto our computer and began at the beginning to write music that synched to the images. Doing it this way, we were also able to accurately synch specific sounds (gun shots for instance) and music to specific images or events in the film. One of the most interesting aspects about writing the soundtrack to a silent film is that you can provide not just the music but all the sound.

Q: What version of the film did you watch before doing the project? Did this have any form of soundtrack?

A: We watched the DVD released in the UK by Tartan and wrote our score to that version. There are a few slightly different edits of Potemkin. As I said above, it had a score edited from music by Shostakovich but we deliberately didn't listen to it.

Q: How did you prepare for the project on a practical level?

A: Loading the film onto our computer. And deciding that the score would be for electronics and string orchestra.

Q: When composing the soundtrack, were you aware of the location at which it was going to be viewed? Did the outdoor setting affect your composition?

A: We were aware of it and the idea of Trafalgar Square as a space for political protest. The only direct influence this had was that in the famous 'Odessa Steps' sequence, the music we wrote was a song called 'After all' which referred to the Iraq war and its aftermath which had been the focus of political protest in the Square.

We also invited the theatre director, Simon McBurney, to stage the event so that the film had a historical and political context within the Square and the audience were drawn into the film.

Q: Did you have any control over the sound system that was used? Did you know this prior to your composing? Did it affect your decision making process?

A: We insisted that our regular sound engineer mix the live sound in conjunction with an engineer who specialized in amplifying orchestras. I think they specified the sound system.

Q: What size orchestra did you have in mind? Have you had to change and adapt the soundtrack to fit different orchestra sizes/compositions/viewing locations?

A: We decided to write for a string orchestra of around 25 players because we like the combination of strings and electronics. When we performed it in 2006 at the Swan Hunter shipyard in Newcastle we had the Northern Sinfonia who had almost twice as many players, giving a richer sound.

Q: Which performance do you feel has best reflected your artistic intention? Why?

A: Having premiered the work in Trafalgar Square, we realized that a location could give a wonderful historical resonance. In the Newcastle performance we were in a shipyard about to close with a battleship (in the process of being renovated) on the River Tyne behind us. In the Dresden performance, the communist recent history of the city resonated with the work. We always have a spoken prologue, sometimes with accompanying images, to relate the film to the setting and to give historical background. The Dresden performance, where the film was projected onto an enormous DDR-era apartment building with each member of the orchestra sitting in a balcony at the side of the screen, was pretty astonishing.

Q: How did you pick your contributors, such as conductor and orchestra?

A: I had read about and then bought the CD of *My Hertz Brennt*, a work by the German composer Torsten Rasch, based on the music of Rammstein, and suggested him as our orchestrator. I emailed his producer, Sven Helbig, who was a co-founder of the Dresdner Sinfoniker who had premiered *Mein Herz Brenn*t and so Torsten, Sven and the Sinfoniker became collaborators on the project. Sven and Torsten suggested the composer Jonathan Stockhammer who has subsequently conducted all performances.

Q: There are strong moments of sync between the visuals and sound? How did you choose which moments to mark sonically and which to let pass?

A: When the film seemed to require samples of sounds as much as music, we added those sounds. We wrote a piece of music inspired by the engine of the battleship. And of course the music and sound were synched accurately to the film.

Q: It is a very full soundtrack with limited moments of silence; what led you to make this decision?

A: There is almost continuous action in the film. The film has its own rhythm which we tried to follow.

Q: There are moments in the film that sound in-line with your previous work and your musical signature, and moments that are quite different. How did you make these choices?

A: I think we have a wider range of styles than most people realise!

Q: How did you synthesise image and sound? How did you understand the relationship?

A: I think I've already answered this.

Q: Were you conscious of the issue of manipulation? How did you approach this issue? Did you consider how your score might interact with Eisenstein's intentions? How do you see this relationship?

A: We read somewhere that Eisenstein had wanted a new score for each new decade and that made us feel happy that we were the latest. We think that it is a film about freedom not communism and so were inspired to write moments of lyricism not just anger.

Q: There are moments of singing and voices in the soundtrack. How did you choose when and how to use/create these moments?

A: In the main they followed the scenes in the film but also the captions on the screen. The song, 'No Time for Tears', for instance, is vocalising the speeches you see on the screen at that point.

Q: Do you plan on doing any more soundtracks?

A: We don't plan not to!

Neil Tennant (Pet Shop Boys)

Notes

1. S. M. Eisenstein, V. I. Pudovkin, and G. V. Alexandrov, 'A Statement on Sound' (1928), in *Film Sound: Theory and Practice*, ed. Elisabeth Weis and John Belton (New York: Columbia University Press, 1985), 84.
2. Ibid.
3. Peter Larsen, *Film Music* (London: Reaktion Books, 2005), 82.
4. Tatiana Egorova, *Soviet Film Music: An Historical Survey* (Amsterdam: Harwood Academic Publishers, 1997), 4.
5. Siegfried Kracauer, *Theory of Film: The Rehabilitation of Physical Reality* (Oxford: Oxford University Press, 1960).
6. Egorova, *Soviet Film Music*, 4–5.
7. Ibid., 5. Denise J. Youngblood has described how Lenin's New Economic Policy (NEP) which started in 1919 led to the reopening of private theatres and the beginning of foreign films being shown, particularly in Moscow. Denise J. Youngblood, *Soviet Cinema in the Silent Era: 1918–1935* (Austin: University of Texas Press, 2014).
8. The interview with the Pet Shop Boys was conducted via email. Many thanks are given for the time and energy they offered this chapter.
9. For a full discography please see 'The Pet Shop Boys', *Discogs*. http://www.discogs.com/artist/7552-Pet-Shop-Boys, accessed 15 March 2015. The albums released by the Pet Shop Boys cover studio albums, live albums, remixes and compilation albums amongst other releases.
10. Mark Butler, 'Taking it Seriously: Intertextuality and Authenticity in Two Covers by the Pet Shop Boys', *Popular Music*, vol. 22, no. 1 (2003), 4.
11. Ibid.
12. Ibid.
13. Fred E. Maus, 'Glamour and Evasion: The Fabulous Ambivalence of the Pet Shop Boys', *Popular Music*, vol. 20, issue 3 (October 2001), 381.
14. D. Travers Scott, 'Intertextuality as "Resonance": Masculinity and Anticapitalism in Pet Shop Boys' Score for *Battleship Potemkin*', *Music, Sound and the Moving Image*, vol. 7, no. 1 (Spring 2013), 55.
15. Ibid., 63.
16. 'The Pet Shop Boys: *Battleship Potemkin*', *Drowned in Sound*. http://drownedinsound.com/releases/4243/reviews/13261-, accessed 15 March 2015.
17. Bill Nichols, '*Battleship Potemkin*: Film Form and Revolution', in *Film Analysis: A Norton Reader*, ed. Jeffrey Geiger and R. L. Rutsky (London: W. W. Norton, 2005), 161.
18. David Bordwell, *The Cinema of Eisenstein* (London: Routledge, 2005), 69.
19. Ibid., 78.
20. Scott, 'Intertextuality as "Resonance"', 55.
21. Ibid., 54.
22. Shelley Cobb, 'Adaptable Bridget: Generic Intertextuality and Postfeminism in *Bridget Jones's Diary*', in *Authorship in Film Adaptation*, ed. Jack Boozer (Austin: University of Texas Press, 2008), 281.
23. Ibid.
24. Robert Stam, 'Beyond Fidelity: The Dialogics of Adaptation', in *Film Adaptation*, ed. James Naremore (New Brunswick: Rutgers University Press, 2000); Cobb, 'Adaptable Bridget', 283.

25. Robert Stam, *Literature and Film: A Guide to the Theory and Practice of Film Adaptation* (Malden, MA: Blackwell, 2005), 45–6.
26. 'The Pet Shop Boys: *Battleship Potemkin*', *Drowned in Sound.* http://drownedinsound.com/releases/4243/reviews/13261-, accessed 15 March 2015.
27. Sarah McDonnell, 'Pet Shop Boys – Battleship Potemkin', *Music OMH*, 5 September 2005. http://www.musicomh.com/reviews/albums/pet-shop-boys-battleship-potemkin-ost#uyW7OuxqrT4MFuTU.99, accessed 15 March 2015.
28. 'Pet Shop Boys: *Battleship Potemkin*'.
29. Egorova, *Soviet Film Music*, 7.
30. Scott, 'Intertextuality as "Resonance"', 64.
31. Maus, 'Glamour and Evasion', 382, 383.
32. Ibid., 386.
33. It is important to note that while many factors could influence the reviewer's reaction to a live performance, whether this be the weather, location, venue or sound system, amongst others, this chapter will focus solely upon their engagement with a particular artistic rhetoric.
34. Maddy Costa, 'Silent Classic Given Soundtrack for Today', *The Guardian*, 13 September 2004. http://www.theguardian.com/uk/2004/sep/13/film.arts, accessed 15 March 2015.
35. Paul Bond, 'Failing Eisenstein: The Pet Shop Boys' New Score for Battleship Potemkin in Trafalgar Square', *World Socialist Website*, 23 September 2004. http://www.wsws.org/en/articles/2004/09/pkin-s23.html, accessed 15 March 2015.
36. Ibid.
37. Ibid.
38. Neil Tennant 'Battleship Potemkin Makes Us Strong', *The Guardian*, 1 February 2008. http://www.theguardian.com/film/filmblog/2008/feb/01/battleshippotemkinmakesusstrong, accessed 28 April 2015. The original reference to Eisenstein's preference for the soundtrack being rewritten is hard to discover, though this idea has been attributed to him on several occasions; *Battleship Potemkin* Press Release, BBC, 9 September 2009. http://www.bbc.co.uk/pressoffice/pressreleases/stories/2009/09_september/09/battleship.shtml, accessed 28 April 2015. Other examples of musicians who have created a score for *Battleship Potemkin* include, but are not limited to: Nikolai Kryukov (1950), Eric Allaman (1986) and the Club Foot Orchestra (2005).

Bibliography

Battleship Potemkin Press Release, BBC, 9 September 2009. http://www.bbc.co.uk/pressoffice/pressreleases/stories/2009/09_september/09/battleship.shtml, accessed 28 April 2015.

Bond, Paul, 'Failing Eisenstein: The Pet Shop Boys' New Score for *Battleship Potemkin* in Trafalgar Square', *World Socialist Website*, 23 September 2004. http://www.wsws.org/en/articles/2004/09/pkin-s23.html, accessed 15 March 2015.

Bordwell, David, *The Cinema of Eisenstein* (London: Routledge, 2005).

Butler, Mark, 'Taking it Seriously: Intertextuality and Authenticity in Two Covers by the Pet Shop Boys', *Popular Music*, vol. 22, no. 1 (2003), 1–19.

Cobb, Shelley, 'Adaptable Bridget: Generic Intertextuality and Postfeminism in *Bridget Jones's Diary*', in *Authorship in Film Adaptation*, ed. Jack Boozer (Austin: University of Texas Press, 2008), 281–304.

Costa, Maddy, 'Silent Classic Given Soundtrack for Today', *The Guardian*, 13 September 2004. http://www.theguardian.com/uk/2004/sep/13/film.arts, accessed 15 March 2015.

Egorova, Tatiana, *Soviet Film Music: An Historical Survey* (Amsterdam: Harwood Academic Publishers, 1997).

Eisenstein, S. M., V. I. Pudovkin, and G. V. Alexandrov, 'A Statement on Sound' (1928), in *Film Sound: Theory and Practice*, ed. Elisabeth Weis and John Belton (New York: Columbia University Press, 1985), 83–5.

Kracauer, Siegfried, *Theory of Film: The Rehabilitation of Physical Reality* (Oxford: Oxford University Press, 1960).

Larsen, Peter, *Film Music* (London: Reaktion Books, 2005).

Maus, Fred E., 'Glamour and Evasion: The Fabulous Ambivalence of the Pet Shop Boys', *Popular Music*, vol. 20, no. 3 (October 2001), 329–93.

McDonnell, Sarah, 'Pet Shop Boys – Battleship Potemkin', *Music OMH*, 5 September 2005. http://www.musicomh.com/reviews/albums/pet-shop-boys-battleship-potemkin-ost#uyW7OuxqrT4MFuTU.99, accessed 15 March 2015.

Nichols, Bill, '*Battleship Potemkin*: Film Form and Revolution', *Film Analysis: A Norton Reader*, ed. Jeffrey Geiger and R. L. Rutsky (London: W. W. Norton, 2005), 158–77.

Scott, D. Travers, 'Intertextuality as "Resonance": Masculinity and Anticapitalism in Pet Shop Boys' Score for Battleship Potemkin', *Music, Sound and the Moving Image*, vol. 7, no. 1 (Spring 2013), 53–82.

Stam, Robert, 'Beyond Fidelity: The Dialogics of Adaptation', in *Film Adaptation*, ed. James Naremore (New Brunswick: Rutgers University Press, 2000), 54–76.

Stam, Robert, *Literature and Film: A Guide to the Theory and Practice of Film Adaptation* (Malden, MA: Blackwell, 2005).

'The Pet Shop Boys: *Battleship Potemkin*', *Drowned in Sound*. http://drownedinsound.com/releases/4243/reviews/13261-, accessed 15 March 2015.

'The Pet Shop Boys', *Discogs*. http://www.discogs.com/artist/7552-Pet-Shop-Boys, accessed 15 March 2015.

Youngblood, Denise J., *Soviet Cinema in the Silent Era: 1918–1935* (Austin: University of Texas Press, 2014).

9

Multiple Soundtrack Versions on DVD: Scoring Modern City Life and Pastoral Countryside

Christopher Natzén

Sunrise: A Song of Two Humans (1927, F. W. Murnau) begins with a train leaving the station followed by a series of juxtaposed shots of more rapidly moving trains, a passenger ship leaving the harbour, and a day at the beach before the sequence ends with an overall shot of a peaceful village. The citizens of the city are leaving town for their summer holidays, and the passing images contrast the hectic modern city life with the calmness and stillness of the countryside. This is further highlighted on the two different soundtracks that can be found on the DVD version of the film released by EUREKA! in their Masters of Cinema series in 2005. The DVD comes with the original Movietone soundtrack produced in 1927 and with a new score written by Timothy Brock and performed by the Olympic Chamber Orchestra. Both soundtracks convey the feeling that the city people are overtaking the peaceful countryside defined by its calmness, but they approach the above sequence slightly differently and thus give two alternative beginnings and two alternative suggestions of how to read the film emotionally. The original score also utilizes sound effects in a rudimentary manner characteristic of early sound film in order to augment the impact of certain aspects of the visual images. In comparison, the new soundtrack uses only music to comment on specific features in the image.

The aim of this chapter is to discuss the radically different experiences of the same film which are made possible through the booming industry of multiple soundtrack versions, providing historically accurate scores as well as re-imagined new versions. Different scores are often distributed and accessed through DVDs that contain multiple versions of the soundtrack. With several choices the viewer is to a great degree constructing his/her own cinematic event in line with Rick Altman's concept of 'cinema as event' where film is viewed with one soundtrack within the walls of the cinema theatre as a self-contained structure around which everything revolves.[1] In other words, the different experiences of a film are dependent on an active choice made by the viewer who decides on which score and which soundtrack to use. As Barbara Klinger explains this, the audience 'does not exist in one stable location in relation to the flux of historical meanings around a film',[2] but is rather in a constant change of positions in relation to a specific film and to the film medium at large. These positions are influenced by new technology, and in this chapter I want to focus specifically on sound technology.

In the following this is illustrated through an analysis of how the two versions of *Sunrise* present slightly different experiences of the conceptual pairing of modernity within the city versus the pastoral life of the countryside. Differences in scoring modern city life will be further demonstrated with the two new scores made for *Underworld* (1927, Josef von Sternberg) on the Criterion Collection DVD released in 2010.

Sunrise: A Song of Two Humans

The original Movietone soundtrack for *Sunrise* projects a gloomy and dull feeling overall while the new score evokes a sadder sentiment. The opening sequence in the Movietone version is built up by starting the music in the credits when 'The Cast' is shown (before this the credits have been shown in complete silence). The music that starts when the participating actors' names are shown instantly hints that this will not be a cheerful story as the music slowly emerges from the preceding silent black and white text signs. This is carried over into the opening train scene where also a sound effect is heard illustrating the leaving train. The music becomes directly associated with the modern world of the city which is further illustrated in the subsequent shots of fast-moving trains and a boat leaving the harbour. This means that the juxtaposed shot of a day at the beach in fact has the character of a 'hectic day at the beach', in sharp contrast with the music that accompanies the first shots of the countryside. The music is here lighter and pastoral giving a calm and relaxed feeling. It is to this location that the city people are heading. Experienced together, the visuals and the soundtrack present the filmic theme of a meeting between the modern city life and a historical rural life, a meeting out of which nothing good can come.

The music on the new soundtrack is different in character as well as constructed in another way in relation to the events as presented in the images. It starts from the very beginning of the opening credits but stops completely before the first shot of the train. This way the themes of the score that have been presented are paused in order to let everything sink in with the audience. Furthermore, the music that accompanies the credits evokes more of a sad feeling than a gloomy one. The ensuing train scene is also approached differently. Where the original maintains the monotony of the gloomy music, this score, as it is starting afresh, has a fateful character that step by step is intensified as the train leaves the station. The music then follows over into the shot sequence of trains, boat and beach but here in some instances the music expresses a happy feeling, seemingly describing the positive emotions associated with escaping the hectic everyday life of the city for a trip to the countryside. The sadness of the music is still present during the shots of the countryside but this time it is played calmer and softer.

In comparison, the two scores set up the story to develop in two different directions. The original Movietone score seems to sympathize with the old-fashioned ways of pastoral life and underlines that the modern ways of the city will abruptly destroy the happy and calm life of the countryside. Modern life is treated as a threat to the existing rural life. The new score, on the other hand, with its slightly lighter character, is more sympathetic to modern city life. It is saying the same thing as the original score and that this is something unavoidable. But it is said with a different angle underlining the inescapability of the automatized modern city life and how it will affect the countryside.

Using different approaches, both scores for *Sunrise* underline the conceptual binary pair of city and countryside which is emphasized in an early scene where a woman of the city – a typical vamp character – tries to convince a husband to kill his wife. The woman describes the wonderful and laissez-faire style of life in the city. This is visually exemplified through a series of shots of city life with its passing masses of moving people, cars and popular jazz bands. The Movietone soundtrack comments on the images by providing a cacophony of city sounds and emphasizing the jazz band thus underlining that this music is something particular to the city. The new soundtrack does a similar thing but gives the overall music a radio quality throughout the sequence. Also here jazz is presented as music from the city but in the latter score this is connected to the radio as a sound carrier that popularizes this type of music in the 1920s, that is, the time period when this film is set.

Both scores emphasize the importance of the new recording technologies for music production and distribution that began to develop at the end of the nineteenth century. Following the turn of the century, the music and sound landscape became increasingly mechanized with gramophones, phonographs and self-playing instruments.[3] Added to this development was radio with its music programmes that became widely popular among the listening public during the 1920s, increasing the presence of mediated music and sounds which were widely associated with modern city life and its inhabitants.[4] Jazz with its dances of one-step and especially foxtrot became associated with the rhetorical figure of 'all that jazz'. The term jazz was widely used and stood in for a whole set of features connected to urban lifestyles. In this way, heard in public places and bought on record, jazz also became popularized as it was easily accessible. This increased mechanization of music was contrasted with a utopian pastoral ideal. As Caryl Flinn has demonstrated, there existed in the United States a utopian ideal of a collective cultural identity realized through music which was believed to have its roots in pastoral life.[5] *Sunrise* sets these two positions against each other, and this can also be heard played out on the two different soundtracks. Approximately ten minutes into the film two elderly women are commenting on the husband's behaviour leading into an idyllic flashback sequence where husband and wife live happily. The Movietone soundtrack illustrates musically what these two diverging positions mean as the flashback scene is scored with light music underlining the brighter and utopian life prior to the arrival of modern city life.

If the original Movietone track clearly divides the two worlds, the new score is filled with dissonances during the flashback sequence as if the past already expresses foreboding about the future. Given the radio quality of the city sequence in the newer score, this illustrates another aspect associated with radio. The technology of radio signalled a shrinking world as it was seen as being a part of new communication techniques together with other transmitters like the telegraph and the telephone that all were developed during the second half of the nineteenth century. Throughout the nineteenth century, there had been a constant progression towards abolishing the restrictions set by space and time, first by the transportation of actual people, later their messages, and finally their voices.[6] Read this way, the new score illustrates that life was not just happy but truly utopian as radio and other similar technologies made life increasingly global.

On the other hand, the Movietone soundtrack also illustrates that while the sentiment of a shared cultural common ground must have been strong, it was at the same time filled with contradictory feelings. The step from being able to share events and impressions only with your nearest neighbour, or in the bigger cities with people in your own part of town, to being capable of giving expression to incidents, experiences and feelings, mediated through widespread radio broadcasting and telephone must have been momentous, as suggested by the diverging emotions expressed in the contrasting gloomy and happier music during several sequences in *Sunrise*.

Underworld

If *Sunrise* is about the clash between modern city life and pastoral countryside, *Underworld* deals solely with the modern city. Specifically it is about what city life does to the people who live there.

This silent feature was at first shelved by Paramount and only released on one single screen in New York. Although the production company did not believe in the film, the audience appreciated it and it became something of a success story. The DVD produced in 2010 contains two scores, one written and performed by Robert Israel and one made by and performed by the Alloy Orchestra. These two scores approach the film in slightly divergent ways. The Israel score brings the work of Max Steiner to mind, while the Alloy score is more fateful as if underscoring the low life and contrasting this with another kind of city life in brighter colours, a life that all characters in the film seem to be unable to reach.

This other city life is given the same role as the countryside in *Sunrise* with the difference that it is not destroyable and there is the suggestion that through a change of life the characters of the lower parts of the city may be allowed to access this ideal life. It is a backdrop that is both there and not there in the film. The only visual connection to this other city is a cat looking at some milk. In this instance the score gives up its fateful character for an instant and is momentarily replaced by a solo harp. The first score also contrasts the split city as downhearted music is replaced by more active and positive music when the escape is planned.

The new communication technologies in the modern city, like the sound film, the radio, the car, and the airplane all represent technology that is seen as systemic. That is, they are for their function not only dependent on technology: many organizational, political, social and cultural aspects need to be developed and incorporated alongside the new technology for it to reach its full potential, thereby overlapping one system of technology with another.[7] However, this creates instability within the new technology as well, as it is 'embedded in social and institutional practice'.[8] In this sense, new technology functions 'as translations of the elusive yet structured social, institutional, and economic relations that partake in their constitution'.[9] This all plays out in a film such as *Underworld* describing how the people of the city are struggling metaphorically to pass the bridge to the other side of the city and what happens when they fail. The ever more complicated communication technologies also involve an increased control of time. This is illustrated in the first score for *Underworld*. When 'the Law' speaks, the music changes character and is underscored by a march, presenting the words like a metronome. As Ben Singer has argued, modernity describes a constant search for order and rationality.[10]

It is important to underline that these technologies, including the gramophone and the sound film, as illustrated in the scores for *Underworld*, were connected to a specific part of city – the underworld – contrasting it with a utopian ideal of a different, better life. Furthermore, positive perceptions of new communication techniques mixed with pre-war ideas about a commerce or entrepreneurship unrestricted by borders, which together created a feeling of almost absolute trust in technical innovation as a requirement for progress. Although this faith received a blow during the First World War, it was still present during the 1920s.

However, experiences linked to new communication techniques are not necessarily connected to economic development. If the capitalist system as such could be questioned, it was not as obvious to question the new communication techniques. Rather, these were seen as a way out of misery and towards progress, and as a solution to many seemingly unsolvable problems as well as possible sources of revenue. But as both scores for *Underworld* exemplify, this also means that the goal of an ordered society must be placed in opposition to a chaotic society. To be able to do this, an imaginary threat against the other must be created as is done in the scores for this film, which both have a dark character.

This is spelled out in *Sunrise*'s Movietone score as well. *Sunrise* is an early example of sound film, a kind of film closely connected to the development of radio and telephone technologies, and to the urban life associated with these sound technologies. Sound films had developed alongside other communication innovations, and can be seen in this regard as the last and final innovation from the turn of the century to reach its fulfilment. In hindsight this gives many films from the period like *Sunrise* and *Underworld* visually a rather ambivalent expression, signalling both acceptance of modern life and reluctance to let go of old values that were fast disappearing in society at large.

Sound cues

I have focused on the actual music scores but the soundtracks for *Sunrise* and *Underworld* also exemplify what can be achieved with a sound film's soundtrack. The different tracks discussed emphasize the importance of sound effects – the Movietone score for *Sunrise* by using sound effects and all other scores discussed in this chapter by opting not to use them. Film has in most cases always been screened with some kind of sonorous accompaniment – be it in the form of a lecturer, sound effects, accompanying music, or for that matter, comments from the audience. During the 1910s these sounds and music were assimilated into the diegesis. Increasingly the aim was that sound effects and music should 'disappear' and be perceived unconsciously, or be 'unheard' to use Claudia Gorbman's notion about classic narrative cinema.[11] This aim to make the sonorous accompaniment 'vanish' resulted in more and more complicated ways of setting sound and music to film of which the *Sunrise* Movietone track is an early example made possible by a relatively new medium as synchronized sound film.

If the new scores for *Sunrise* and *Underworld* use the music mainly to give an overall feeling, the original soundtrack for *Sunrise* uses both music and pure sound effects. Through this combination the soundtrack is constantly commenting on central aspects of the visuals, for example with the sound effect given to a suddenly emerging horse when the husband hides the reed bundles that will keep him floating. The new soundtrack for *Sunrise* makes a similar comment but solely through music.

Technically the different scores thereby work differently: whereas the original Movietone soundtrack, made for a sound film, more persistently commented on visual aspects, the newly written scores almost exclusively give an overall feeling to the unfolding events. Seen together the soundtracks illustrate what Alan Williams has pointed out in relation to the soundtrack for the early sound film *White Shadows in the South Seas* (1928, W. S. Van Dyke). Williams argues that one 'noteworthy aspect of the accompaniment to *White Shadows in the South Seas* is how the various materials are associated with fictional elements, given that they are character- or even situation-specific'.[12] This happens to a much higher degree in the original Movietone soundtrack for *Sunrise* compared with the other soundtracks analysed in this chapter. The reason behind this is partly due to the fact that the original score is more economical in its structure, working with smaller themes, whereas the others are more time-consuming forcing the score to work on its material before it can come to an appropriate musical closure.

Conclusion

To sum up, the different scores of *Sunrise* and *Underworld* show in different ways the clash between various parts of modern city life and pastoral countryside. Visually both films were produced in 1927 while only one of the scores discussed here was made at this time. The different creation years of the scores are evident as the Movietone score from 1927 incorporates several new sound film techniques.

A typical example is the husband's 'screams' in *Sunrise* when he is searching for his wife after the storm. Every single scream is emphasized in the Movietone track with an oboe, an instrument that on this occasion, after it has been established diegetically, is allowed to be heard 'off-screen', creating a feeling of an endless and desperate search. The new version of the soundtrack resolves the same scene by providing an overall feeling of desperation without emphasizing the screams. Similar differences can be heard the first time the woman from the city is seen whistling after the husband at the beginning of the film. The 'whistling' is continued throughout the sequence, emphasizing the husband's emotions that are clearly illustrated visually. The new score only makes a similar comment when the woman is clearly seen in the image whistling before the score drops this idea entirely.

It is as if in the Movietone soundtrack for *Sunrise* we are witnessing 'the ever-closer relationship between the sensuous and the technological', as Sara Danius wrote in her study on the configuration of seeing and hearing in the modernist period.[13] These sonorous cues were underlined with music, creating a feeling of a unified medium. The new 'silent' scores for *Sunrise* and *Underworld* bridge the gap between the sensuous and the technological, creating a continuous flow of exchange between music and images.

Notes

1. Rick Altman, 'General Introduction: Cinema as Event', in *Sound Theory, Sound Practice*, ed. Rick Altman (New York: Routledge, 1992), 2–4.
2. Barbara Klinger, 'Film History Terminable and Interminable: Recovering the Past in Reception Studies', *Screen*, vol. 38, no. 2 (Summer 1997), 114.
3. James P. Kraft, *Stage to Studio: Musicians and the Sound Revolution, 1890–1950* (Baltimore and London: Johns Hopkins University Press, 1996), 1–2.
4. Ibid., 63.
5. Caryl Flinn, *Strains of Utopia: Gender, Nostalgia, and Hollywood Film Music* (Princeton, NJ: Princeton University Press, 1992), 22.
6. Mats Björkin, *Amerikanism, Bolsjevism och korta kjolar: filmen och dess publik i Sverige under 1920-talet* (Stockholm: Aura förlag, 1998), 33.

7. Martin Kylhammar, *Frejdiga framstegsmän och visionära världsmedborgare: epokskiftet 20-tal – 30-tal genom fem unga och Lubbe Nordström* (Stockholm: Akademeja, 1994), 45–6.
8. Sara Danius, *The Senses of Modernism: Technology, Perception, and Aesthetics* (Ithaca and London: Cornell University Press, 2002), 43.
9. Ibid.
10. Ben Singer, *Melodrama and Modernity: Early Sensational Cinema and Its Contexts* (New York: Columbia University Press, 2001).
11. Claudia Gorbman, *Unheard Melodies: Narrative Film Music* (Bloomington and London: Indiana University Press, 1987).
12. Alan Williams, 'The Raw and the Coded: Sound Conventions and the Transition of the Talkies', in *Cinesonic: The World of Sound Film*, ed. Philip Brophy (North Ryde: Australian Film Television & Radio School, 1999), 239.
13. Danius, *Senses of Modernism*, 2.

Bibliography

Altman, Rick, 'General Introduction: Cinema as Event', in *Sound Theory, Sound Practice*, ed. Rick Altman (New York: Routledge, 1992).

Björkin, Mats, *Amerikanism, Bolsjevism och korta kjolar: filmen och dess publik i Sverige under 1920-talet* (Stockholm: Aura förlag, 1998).

Danius, Sara, *The Senses of Modernism: Technology, Perception, and Aesthetics* (Ithaca and London: Cornell University Press, 2002).

Flinn, Caryl, *Strains of Utopia: Gender, Nostalgia, and Hollywood Film Music* (Princeton, NJ: Princeton University Press, 1992).

Gorbman, Claudia, *Unheard Melodies: Narrative Film Music* (Bloomington and London: Indiana University Press, 1987).

Klinger, Barbara, 'Film History Terminable and Interminable: Recovering the Past in Reception Studies', *Screen*, vol. 38, no. 2 (Summer 1997).

Kraft, James P., *Stage to Studio: Musicians and the Sound Revolution, 1890–1950* (Baltimore and London: Johns Hopkins University Press, 1996).

Kylhammar, Martin, *Frejdiga framstegsmän och visionära världsmedborgare: epokskiftet 20-tal – 30-tal genom fem unga och Lubbe Nordström* (Stockholm: Akademeja,1994).

Singer, Ben, *Melodrama and Modernity: Early Sensational Cinema and Its Contexts* (New York: Columbia University Press, 2001).

Williams, Alan, 'The Raw and the Coded: Sound Conventions and the Transition of the Talkies', in *Cinesonic: The World of Sound Film*, ed. Philip Brophy (North Ryde: Australian Film Television & Radio School, 1999).

Part III

Current Practices and New Traditions

10
Edit's Hand: Music to *The Phantom Carriage*

Matti Bye

When I first saw sister Edit gently and tenderly put her hand on the drunken David Holm's soiled coat in the film *The Phantom Carriage* (1921), her hand lingered there for a moment which became an eternity, and this changed my view of the silent film radically. In this lingering gesture her hidden feelings for this man to whom she was, against all reason, so strongly attracted, were suddenly revealed. There was a sensitivity and an expressiveness in Victor Sjöström's artistic direction that before this I did not know existed in silent film. This was art with a visual poetry and a natural creation so subtle and nuanced that I was amazed. Before that I had accompanied as a pianist a large number of silent films but thought that most of the films I accompanied had something superficial and exaggerated in their expression, which often left me unaffected afterwards, even if the films themselves were fascinating with many fine details. In the film *The Phantom Carriage,* I saw something that changed me as a silent film musician because I saw a drama that was told in a timeless style with an empathetic warm tone and with such a careful and personalized voice, which at the same time was clear, individual and distinct. The story and the film experience were not a nostalgic trip but felt of current interest. It was universally human and real. (If you missed that scene in *The Phantom Carriage* you missed something important in early film history – the sensitivity evident in small nuances of imagery.) The movie experience made me henceforth always defend silent film in an overprotective manner against those who thought that early films were only theatrical and excessive. Watch *The Phantom Carriage*! I said euphorically to the uninitiated.

My first assignment as a film composer was to write new music to Victor Sjöström's *The Phantom Carriage* for an ensemble of nine musicians (two violins, viola, cello, clarinet, trombone, harp, percussion and myself on piano and organ). The assignment came from SFI (Svensk Filmindustri, Sweden's national film company). Before this task, I had improvised to silent film for eleven years and had a classical background as a pianist and theatre musician. Much of the music I did as silent film pianist was usually spontaneously improvised directly to a film that I had never seen before and now I had to develop themes and sequences into a complete score to be played by an ensemble. It became necessary to formulate what for years I had instinctively performed.

I had to go back to my first experience of playing to a silent film. This was Sergei Eisenstein's film *Battleship Potemkin* (1925) in a high school auditorium in 1985. The anarcho-syndicalists at this school wanted to show the film during a lunch break for the school's students, and they asked if I could accompany it. I thought it sounded exciting and accepted the assignment. But how do you actually play to a silent film? My Russian piano teacher said, somewhat in passing, that silent film music was structured on diminished triads and the first composer I thought about was Beethoven. Beethoven's piano sonata *Pathétique* Opus 13 was the music I chose to have as a basis to accompany *Battleship Potemkin*. The piano sonata's strong melodies, the dramatic dramaturgy, the abrupt shifts in dynamics between *ppp* and *fff*, the tension-filled diminished triads and the simple chord progressions and cadences should fit perfectly to the silent film, I thought. The big shock came some way into the film: *Battleship Potemkin* is 75 minutes long and the *Pathétique* sonata about 18 minutes! What do I do when I finish playing it? Start all over again? In sheer panic, I randomly replayed a number of bars and began to improvise using the major and minor scale tones of the chords. I began cautiously changing the tempo and slowly I became irresistibly drawn into the film's story which somehow became more important depending on how I chose to play. Suddenly I understood the magnitude of how the music directly and extremely strongly influenced the significance of the pictures. A huge responsibility rests on the pianist, I realized with awe. The choice of music and especially the dynamics of touch decide how the audience experiences the film. I repeated the chords in the *Pathétique* sonata in endless reruns. I had not made any major preparation of the accompaniment, nor had I studied the film's dramaturgy, so the rapid revolutionary editing in the epoch-making scenes dictated that I was hopelessly behind in my musical reactions to the events. Gullibly, I had thought it enough to play some classical pieces to the film. I treated it very lightly. The practical experience was shocking. This was music on-the-spot. I had no experience of free improvisation.

In addition to classical piano lessons in the music programme, I also took lessons from jazz pianist Robert Malmberg and there I had played around with chords and modal jazz improvisation. But for *Battleship Potemkin* I immediately felt that jazz chords and melodics were out of place for this particular film; I wanted the music to have a classical feeling. So, in sheer panic, I used Beethoven chord modules which I elaborated on and instead of jazz scales I used simple major and minor scales (and the more unusual diminished scale, which alternates full and half-tones).

After this tumultuous experience, it was another year before I got the chance to try silent film accompaniment again when the cinemathèque programme's editorial called and asked if I could play for a few silent films at the Filmhuset (Film House) in Stockholm. One of the women who were part of the programme editorial had in fact been (one of the few) in the audience at the screening of *Battleship Potemkin* at the high school and remembered me as the silent film pianist. This time I wanted to be more prepared and so I went to the film library at the Filmhuset. I found almost nothing about silent film music, except one book, which became an inspiration because it was written from practical experience and application of music accompaniment to silent films. It was called *Musical Accompaniment of Moving Pictures* by Edith Lang and George West from 1920. There was quite a lot of theoretical literature on early film music but no musical notes. The projectionist at the cinema lent me Ernö Rapée's collection *Motion Picture Moods for Pianists and Organists* from 1925, a manual with short themes for different moods possible to develop and improvise over, that were meant to be applied directly to film. It was exciting to look at but somehow I never used it in practice. Instead I put together my own album with selected classical fragments and themes from composers who inspired me more: Beethoven, Schubert, Prokofiev and Grieg's melody worlds and chord progressions and Debussy's and Ravel's impressionistic sound worlds and whole tone scales. Satie was also a composer that I used as inspiration for silent film accompaniment, and even modern film composers like Ennio Morricone and Nino Rota. But the big idol and inspiration was Shostakovich. His dramaturgical sense, strong melodies and distinctive modulation art were something to be constantly inspired by when I improvised and composed music for film.

Many years later I got a call from the Circus archive and they told me they had found a hardback leather folder with notes that they thought

I might be interested in as a silent film pianist. These turned out to be conductor Rudolf Sahlberg's original notes from his time as a conductor at the première cinema theatre in Sweden, Röda Kvarn in Stockholm. The folder included full programmes pasted into a fine album with the years inscribed, written with personal and swift handwriting. This stated what themes and pieces they played to each movie première. It was wonderful to read as a history document. But, and here we come to an important detail, it included not a word about the dynamics or where in the movie the theme started or stopped; no cues.[1]

After a few years, I gradually abandoned the obvious classical quotations and instead began to compose my own melodies and improvisation modules. Inspired by rock music and contemporary art music such as that of John Cage, Arvo Pärt and Sofia Gubajdolina, my music became more experimental, and lately the music that I write and improvise to silent film is also mixed with live electronics using a contact microphone on the piano and electronic instruments. So when it was time to prepare for the work to write new music for *The Phantom Carriage* for an ensemble, I had to formulate my own credo. I spent a lot of time determining the instrumentation, which is so important in all film music. But the important and difficult thing in ensemble playing was not *what* music would be played but *how* it would be presented along with the image. Was it even possible to write this down in notes?

I chose classical musicians with no experience of improvising, nor had they ever played to silent film. They expected a comedy and got a melodrama; moreover, the music was notated in a Spartan way without dynamic indications and sometimes without bar-lines. I wanted the written score to serve as an inspiration. So I had to formulate two rules for the ensemble:

1. Eyes on the screen
2. The dynamics are in the movie.

The film is the score, I explained. Everything is there. I became a musical director; we talked about the scenes and the plot and we made interpretations of the characters and read the novel as if we were a theatre group. It is about empathy and compassion, I said. You Anna, you *are* Edit, you should follow her like a ghost with your violin, and every grimace on her face shall surface in the music! And you Ivo, with your trombone you must catch Happy Georg in such a way that we feel sympathy with him. He has a dual nature, as a manic-depressive. He tries to be happy but is deeply grieved and you can be both things at the same time, can't you? These notes are just sketches that will open up doors to the emotions and it is your way of playing it that will make it possible. I need your direct response to what we see on the screen. You should create the tones in the moment so that they emanate from you, and no notes are 'wrong' as long as you play them with empathy for the story. A 'wrong' tone just feels 'wrong' if you play it once; repeat the note that feels 'wrong' and suddenly it has got meaning and content and will no longer be 'wrong', perhaps odd, but not 'wrong'. The music must come from within otherwise the film does not live; we must awaken the dead.

A dynamic play makes the images speak more directly to the viewer on a human and emotional level and can thus also leave room for the audience's free associations. In my work to improvise and compose music for silent film this became perhaps the most important aspect to emphasize in early films, that is, not only to present a historical document but to imbue the characters with life and sometimes let the music tell us something more than what is immediately apprehensible on the screen, that which might be hidden in the story. For example, Edit's revealing encounter with David at the slum station is a key scene for this. A tango played in a scene set in a slum station in Sweden in the early 1900s is probably not correct on a superficial historical plane, but the music instead places the focus on the inner drama that is going on right in front of us. The tango and especially how it is dynamically presented by the musicians tell us something about Edit's psychological condition. This was an interpretation that I made based on the novel of *The Phantom Carriage* by Selma Lagerlöf but it was also perceptible very much in the film narrative because of the film's direction and physical performance of the encounter of the two characters. To choose a tango and to play it the way we did was perhaps a contrasting selection regarding the scene's environment but the way the musicians played it precisely followed Edit's expression and what she experiences emotionally. And that was the important thing. Asynchronous music as René Clair described in his early essays inspired me to think along those lines. (Why do we have to hear how a door closes if we simultaneously see it being shut? Why not replace the realistic door sound with a more interesting sound?)

I wanted to create an emotional and psychological equivalent to the term 'mickey-mousing' – a technique for the musicians to instinctively and spontaneously react to any expression of emotion on the screen as quickly and efficiently as the talented musicians in Hollywood followed the physical gags in the early animated comedies with musical accuracy. We did the same thing but instead put the focus on the synchronization with feelings in the melodrama. In that way we could accentuate the characters' inner lives through the music. The music occupied a psychological place. That demanded a personal commitment of a completely different nature than what the musicians were accustomed to in regular music reading. It required experience and empathy as a musician and the ability to feel the fusion of music and image. (I remember when I was eleven years old and watched a Tarzan movie for the first time, *Tarzan and the Amazons* (1945, Kurt Neumann). The feeling afterwards, when I left the neighbourhood movie theatre at Sibyllegatan in Stockholm and felt that I was still in the movie – I 'was' Tarzan – and then how that feeling slowly disappeared, ran away in the face of reality, the streets and the city beyond. A feeling of merging that the cinema created in its magical world.)

Our starting point in the rehearsal work on the music for *The Phantom Carriage* was then when Edit puts her hand on David. That gesture revealed to me her attraction to him, an impossible earthly love, yet real. I chose to write a tango for their meeting. Not music that was obvious and more realistic in the environment, such as a hymn would have been, but a passionate tango, a distinctive style of music Edit in reality probably would never have known, but I wanted to highlight and show that she felt a hidden and secret passion for David. She felt a sensual attraction to him. And a tango was suitable for this. Perhaps the magical realism in Selma Lagerlöf's story also opened up a door for these musical ideas since the realism was mystical, symbolic, and perhaps you could say subconscious. The ghostly character in *The Phantom Carriage* works more on a psychological level which is about remorse and guilt and how these emotions affect people. Even such feelings could be highlighted with the music, such as the heavy and slow musical part in the reconciliation scene between David and his wife. In that scene the music takes its place in David Holm's bad conscience and his heavy steps on the way to the meeting where an inner decision is taken and has now taken shape in the reconciliation with his wife. Another instance is the café scene where David sits and drinks with his friends and where the music reflects intoxication through constantly having an echo in the piano melody, while a clarinet embodies this through always being a little late, which is musically portrayed by a little lazy touch with a short delay of each note in the melody like a drunk echo in David Holm's head. The musicians played it with an emotionally precise delayed synchronization.

I am often asked by a curious audience what music was played during the silent film era. The music used at that time was largely determined by the musician(s) and the orchestra leader's individual taste in music and the technical level they mastered. During the silent film era the musical artistic choices may also well have been made in passing by the cinema theatre owner who after all in the end was the one who hired the musicians. There is something random and fleeting in the answers about what music actually was used back then. There were so many pianists and so many cinemas. Premières are often satisfactorily documented in terms of their music material but what about all the other screenings? What did the pianist play in the small provincial theatre? In my early quest for guidance to early film music accompaniment, I was more interested in descriptions of *how* the music was played and performed than which pieces were used. How did the improvisation work in practice? Which dynamics did the pianist use in his meeting with the film? What kind of timing and pauses were used? What choices did the musician(s) make in order to follow the dramaturgy of a specific scene? Did they always choose to create a musical counterpart to the environment or did they also focus on the psychological state of the characters? How did the music move forward with the movie? For the issues of dynamics and spontaneous dramaturgical interpretation in the improvisation, there were few, if any, pieces of data. What I soon realized through my own experience as a pianist was that if you play strongly and without dynamism throughout a whole movie it makes the audience insensitive to the details and nuances of the film. The audience becomes inattentive to small but important details in the story which reduces the film's quality and complexity. The soundtrack can take the viewer to new places if it wants but the neutral, uninvolved silent film accompaniment can create a cool distance to the pictures instead of diving into the images and creating alchemy with the film.

A great responsibility therefore rests with the musicians who accompany a silent film. They will strongly affect the audience's viewing experience through their personal interpretation. It is thus also a dramaturgical task that rests on the musician's shoulders. Music is of course also like film based on dramaturgy. Victor Sjöstrom, like many other directors of the silent era, was oddly enough uninterested in what music was played to his silent films.[2] This might have been due to disinterest, lack of time or that he simply did not think it was his job to comment on the musical accompaniment. Victor Sjöstrom seems completely to have trusted conductor Rudolf Sahlberg when he and his family went to the premières of his films at the Röda Kvarn cinema in Stockholm.

So what is silent film music today? It is very much a live art form where newly composed music is performed as film concerts at festivals around the world and with new young pianists who are trained by the more experienced silent film pianists in Pordenone during the workshops that take place in conjunction with Le Giornate del Cinema Muto, and in new DVD releases of the newly restored silent film with an increasing artistic awareness of the important role of music to silent film which mainly has occurred in recent years.

In my view, there is really not much difference in the work and approach whether writing music for silent films or to a modern movie (besides the important point that in the latter work, you are in constant dialogue with a living director). The sometimes fuzzy boundaries that define silent film music and modern film music were sharpened when I worked with the music for Jan Troell's film *Everlasting Moments* (2008). Jan Troell first edited his film to the music of *The Phantom Carriage* (like a temp track) which we later exchanged for the new music I wrote to the scenes. The musicians were the same musicians that I often play with to silent films and who over the years have developed fingertip sensitivity and dramaturgical intelligence. They enter empathically into the movie and the characters in an intuitive and dynamic way. We used the same empathy working on the music for *Everlasting Moments* as we did the first time to *The Phantom Carriage* and naturally we played and made recordings also this time directly to the scenes on a screen in front of us.

Epilogue

I met Aki Kaurismäki at the Midnight Sun Film Festival in Sodankylä in 2003. We had performed the music for *The Phantom Carriage* and he told me as we sat in the smoke sauna after the screening that he was a great admirer of Selma Lagerlöf and had all her collected novels in leather straps at home in a secret cupboard. 'They are all good film scripts', he said. It was the same festival where *The Man Without a Past* (2002) was shown publicly for the first time, and we saw the festival director Peter von Bagh wander around wrapped in a big blanket and in deep conversation outside in the bright birch forest. It was the time of year when the nights never get dark there and I asked Aki cautiously, knowing that he could easily fly off the handle, if his film was inspired by *The Phantom Carriage*. I saw so many similarities especially in the opening, the church bell striking twelve times before David dies, the Salvation Army, etc. So I just had to ask. He smiled his wolf smile, looking furtively at me, took a long drag on his cigarette and said softly in his particular and enigmatic way: 'Don't analyse' (pause; draws on cigarette). 'Just create'.

Translated by Ann-Kristin Wallengren

Notes

1. Ann-Kristin Wallengren uses and discusses this material in her dissertation, *En afton på Röda Kvarn: Svensk stumfilm som musikdrama* (*An Evening at Röda Kvarn: Swedish Silent Film as Music Drama*) (Lund University Press, Dissertation, 1998).
2. Bengt Forslund, *Victor Sjöström: hans liv och verk* (*Victor Sjöström: His Life and Work*) (Bonnier, Dissertation, University of Stockholm, 1980).

Bibliography

Forslund, Bengt, 'Victor Sjöström: hans liv och verk' ['Victor Sjöström: His Life and Work'] (Dissertation, Bonnier, University of Stockholm, 1980).

Wallengren, Ann Kristin, *En afton på Röda Kvarn: Svensk stumfilm som musikdrama* [*An Evening at Röda Kvarn: Swedish Silent Film as Music Drama*] (Lund: Lund University Press, 1998).

11
Scoring Ruttmann's *Berlin*: Musical Meaning in Historical and Critical Contexts

Matthew Malsky

> What have the subtle changes and forms of rhythm in Walter Ruttmann's *Berlin* in common with the trams shown in the film? What have the shots of Montmartre streets in Cavalcanti's *Rien que les heures* in common with the legato-staccato of his cutting? From the viewpoint of rhythm these features are merely carriers of light and shadow, of form and movement. They are no longer objects at all. The visual music of the montage is played in a separate sphere that is parallel to the content.[1]

> I urge the criticism because *Berlin* still excites the mind of the young, and the symphony form is still their most popular persuasion.[2]

When I first became acquainted with Walter Ruttmann's 1927 silent film *Berlin, Symphony of a Great City*, I was taken with its vitality and visual virtuosity. I was fascinated by its rhapsodic and poetic staging of a single day in Berlin from dawn to midnight. I found its muscular representation of trains and factory machinery, all in motion, exciting and its preservation of an historical moment compelling. I was intrigued by Carl Mayer's motivating conception: a 'city symphony'. This film takes Berlin as its focus, but embodies competing principles in its construction. Documentary footage is organized through the application of avant-garde techniques. To create the source footage, Ruttmann shot on location for an entire year, capturing the spaces and activities of Berlin as a photographic archive.[3] Each scene of the pulsating life of the city was collected and tagged as an entry in what Michael Cowen describes as database. Ruttmann then built the film without characters or more than a simple framework – a 'day-in-the-city'. Instead, the film was constructed using the logic of the *Querschnittfilm* (cross-section film). Like cross-sectional illustration in medicine, botany or geology that exposed the internal and the external through simplified visual representation, this film dissected and recorded mass culture. In capturing quotidian life, the film sought to do more than simply document. Ruttmann desired to uncover simultaneously the functional relations of structure and detail, the inner operations and secrets of its subject: the city. Yet in *Berlin*, the penetrating realism of the photographic images was choreographed according to the laws of visual spectacle.[4] It was the rhythmic propulsiveness of Ruttmann's fluid, rapid montage that Béla Balázs dubbed in the epigram at the start of this chapter as visual (or optical) music.[5] For Balázs, this film was characterized by an artistic practice that privileged experimental form and technique. The modern city that emerged in Ruttmann's *Berlin* fascinated and frightened contemporary critics such as John Grierson and Siegfried Kracauer, much as did modern mass culture itself. But despite early critical misgivings, the film's initial reception was overwhelmingly positive, and it has persisted as a singular visual document. It now offers a bundle of contradictions which can be understood in a competing ways: a window into the lost world of Weimar Berlin, a unique and compelling aesthetic object, the quintessence of the genre, or a foreshadowing of fascistic tendencies.

But how do we come to know *Berlin* when we listen as well as watch? As a composer, I have been most struck that these critiques rarely acknowledge the presence of music or consider its function. This is an especially compelling question because the film is predicated on such an elaborate musical metaphor, and also because it has an extensive musical history. A commissioned score was integral to the creation of the film. Composer Edmund Meisel collaborated closely with Ruttmann, and his original music was performed live and in close synchronization with the film at Berlin's Tauentzien-Palast cinema on 23 September 1927. The first showing for the press was successful enough that the theatre owner sanctioned further public performances. It ran for several weeks with Meisel's accompaniment.[6] On 6 May 1928, The *New York Times* reported that screenings of *Berlin* planned in America would include Meisel's score.[7]

Despite its success, Meisel's score was thought 'lost' until recently, and the film was likely often screened without music or with a variety of arbitrarily chosen musical scores until it appeared on VHS tape in the 1980s. However, since the revival in the 1990s of enthusiasm for live music performance with silent film, Ruttmann's *Berlin* has attracted the attention of composers and improvisers who have created new, 'purpose-built' sonic accompaniments for the fixed image track. As a result, it has accumulated a substantial number of quite diverse soundtracks. Undoubtedly, each one has contributed to the experience of this film in a different way. In this chapter, I will consider the function of three different musical scores with the opening sequence of Ruttmann's image track: Meisel's 1927 score, a bit of American jazz that accompanies a fragment of *Berlin* as it appears in a documentary from 1985, and my own, recent string quartet for the first *Akt*.[8] I will consider each of these musical examples in combination with the image track as historically and socially situated, but also as interchangeable elements in a commutation test. I will consider the combinations both synchronically as elements in an evolving history of the work, as well as diachronically as paradigmatic elements whose features shift our attention and focus, and sharpen the ideological message of the film.

Watching the arrival

In the oft-analysed opening sequence of the film, the viewer is brought into the city. As the sun rises, a train speeds across the countryside passing through an increasingly urban landscape to arrive finally at Berlin's Anhalter railroad station. After the credits but before the train appears, we see an abstract pattern created by gently undulating water, which viewed close up and presented without borders represents 'undifferentiated matter [...] a primordial state of purity'.[9] This dissolves gradually into a hand-drawn animation thoroughly reminiscent of earlier abstract experiments – an intertextual reference to Ruttmann's 'absolute' films, *Opus I–IV*. Composed of pulsating horizontal lines and circles that rotate circularly through the frame, this striking animation in *Berlin* prefigures the shapes and motions of the train, and through acceleration, sets its tempo. The animation ends with falling diagonal lines, anchored to the screen's bottom corners, which dissolve to reveal warning gates closing at a train crossing. The animation has thus moved us from the reductive and schematic to the photographic world. But it also emphasizes the film's preoccupation with movement and the formal play of visual elements. Circles and lines in motion continue – in the form of photographed train wheels and tracks, telegraph poles and wire – to predominate throughout this sequence. As if recapitulating a historical or even evolutionary path from a watery genesis to urban modernity, this sequence introduces the train as the quintessentially modern machine, first defined through its abstract and formal properties. It arrives to overwhelm us visually, looming large and moving fast. The opening sequence winds down as a sign announces 'Berlin' with an arrow and a distance indication of 15 kilometres. The landscape changes from rural to increasingly urban with the introduction of modest workers' cottages, followed by large tenement blocks, and then the industrial rail yard. A rail switch and signal mark our entrance into the station itself as we follow the tracks to the end of the line. The station's 'Berlin' sign and the steam released by the engine's brakes signal the end of the journey. We've arrived in the city.

There are two competing visual strategies in this opening. The first forgoes reference to diegetically coherent space. For example soon after the animation, we see close-ups of train wheels, pistons, tracks, telegraph poles, wires and bridge trestles separated by sixteen quick jump cuts. We rush along at breakneck speed both within the frame and between shots, viewing isolated details of a succession of man-made elements. The world is presented through segmentation without match cuts or continuity editing. Rather, metric editing predominates, with each image allotted the same number of frames. This strategy emphasizes rhythmic, spectacular and formal aspects of the visual track, and returns periodically through the opening sequence – a close up of train car couplings, for example – to interrupt the second strategy. A different type of visual material predominates during the later part of the sequence: middle or long shots establish a viewing position within (or alongside) the train. Continuity cutting and analytic editing maintain spatio-temporal coherence.[10] We're watching the landscape pass on the train's right side in accordance with the 180-degree rule. We're situated in conventionally understandable space and time. As a result, there are competing subject positions presented by the image track: an impossible, super-human perspective on the train itself versus a comfortable seat in the coach.

Meisel's *Berlin* as cross-section, 1927

I'd like to suggest three ways in which elements of Meisel's musical soundtrack combine with this image track to privilege the first strategy, encouraging viewers to identify with the train itself.[11] Viewers experience this journey as if the train's commanding power, energy and presence is the film's message. This is accomplished through a combination of mimetic sound effects, musical elements that operate as metaphors, and moments of concordance between sound and image tracks. In an April 1927 article in *Ufa-Magazin*, Meisel described his work on *Berlin* as

> a conglomeration of all the sounds of a cosmopolitan city. For the first time in an ideal manner, film and music are going hand in hand from the outset, a work is generated collectively by director and composer. I intend to create a symphony of our capital city, which – detached from the film – should also be performed in the concert hall, and moreover I am writing it in the rhythm of our times through the use of completely new means and instruments.[12]

Meisel sought to exceed the common film music practice of the day in multiple ways. *Autoren illustration*, or 'author's illustration' was similar to a compiled score, in which pre-existing music was assigned to the image after a film had been completed. The composer was never a participant in the artistic planning or production of the film.[13] By contrast, Meisel composed an *Original komposition* or 'special' score, representing an ideal collaboration between filmmaker and composer from conception through completion of the film.[14] As Nora Alter notes, Meisel conceived of his score as an assemblage of characteristic cosmopolitan sounds and expected that his urban audience would recognize their quotidian sonic environment in his music.[15] Every sound in Meisel's score for this opening is dedicated to the portrayal of the train. We hear the train through the music. Train bells are appropriated as musical instruments. Whistles are duplicated by sustained perfect fifths in wind instruments. Chattering, noisy percussion, a thumping bass drum and chugging low brass reproduce the sounds of the locomotive's engine and wheels. Additionally, Meisel created a new sound effects desk to supplement the musical instruments with additional noises, 'a futuristic device that was capable of imitating claxons, trains, motor-cars or shrieks, either *pianissimo* or *fortissimo*'.[16] Furthermore, as described by Meisel, the instrumentalists were positioned around the auditorium for spatial effect 'in an experiment to attain the illusion of totality'.[17] Each of these musical sounds operated as an indexical signifier, linking a sound to a source present in the image. Characterized by their mimetic, referential qualities, these noises and musical sounds by convention and practice stood in for the actual sounds of the train.

The conventions for reproducing the sound of trains in orchestral music were well established by 1927. Precedents include works by Hans Christian Lumbye, Jacques Ibert, Vincent d'Indy and Johann Strauss (the younger). Arthur Honegger's modernist work *Pacific 231* was popularly understood to represent a steam locomotive. This association was likely reinforced by Honegger's other 'train' music for Abel Gance's *La roué* (France, 1923) which Lucy Fischer called the single dramatic film that could 'stand as both harbinger and emblem of cinematic modernity', and featured at its opening a rapid montage sequence of a locomotive.[18]

At a higher level of abstraction, Meisel's use of the orchestra, his musical phrasing, textures and formal plan participate as Peircean symbols, suggesting mechanization and the subject relations of modernity. Meisel's score called for an orchestra of 75 musicians with a full battery of percussion (including anvil and steel sheets), a harmonium tuned in quarter-tones, and an additional jazz ensemble for the final night-life sequence – truly a work that could only be performed with the resources available in a great city. The choice of orchestra as the vehicle to realize this soundtrack is meaningful, even beyond its sonic power. R. Murray Schafer has compellingly described the orchestra as the archetypal instrument of musical modernity, a symbol of industrialization.[19] In its social organization, it reproduces the hierarchies of a factory. The conductor and musicians are management and labour. In Schafer's Fordist view of the orchestra, each musician performs as a specialized though ultimately replaceable element in an industrialized form of mass production. Meisel has composed a texture consisting of complex, rhythmically interlocking musical lines – an intricate homophony that sounds like the synchronized moving parts of a machine. This type of music may only be realized through the social organization of the orchestra, and stands in distinction to the communal, cooperative time-keeping and synchronization of other ensembles, such as chamber music, for example. Finally, the musical form is a metaphor for modernity. The short phrases of irregular length combine into a fragmentary, non-developmental, 'modular' structure, which is appropriate to support the repetitive, but discontinuous mechanical actions of modern trains or films.[20]

Finally, it is instructive to consider those moments of concordance when the music and image tracks move in unison. Advance press notices for *Berlin* revealed that Meisel would be utilizing a relatively new invention, Carl Robert Blum's *Musik synchromometer*, to secure tight synchronization between music and picture.[21] Blum's invention was an electrical device which, when coupled to the film projector, enabled predetermined tempi to be reliably set in live performance. This metronomic information, together with precise indications in the written score, indicates that Meisel presumed a high degree of coordination between his music and the image track. This technological advance made synchresis possible, the forging of an immediate, repeatable and necessary relationship between what is seen and what is heard.[22] Meisel and Ruttmann carefully crafted the synchronization of formalist visual material and musical phrases. For example, just after the photographic train appears on screen each of the sixteen quick cuts discussed above coincides exactly with a beat of music. Thus, repetition and phrasing in the music organizes the steady 'beat' of the visual track perceptually into groupings: 2 + 2 beats as a unit is followed by an irregular 3-beat phrase, which is followed by two different 4-beat phrases. This type of coordination was newly possible with Blum's new synchronizing technology, and it must have been striking to audiences in 1927. Later in the sequence and absent the metric editing, the music provides continuity across the visual cuts in the sections of the image track organized with continuity cutting and analytic editing. Yet, the presence of the train established early by image and music in tight synchronization is continued throughout the sequence by just the music.

The critical responses to the première with Meisel's music revealed a contradiction. On the one hand, a sympathetic contemporary of the film described its 'wish to give pleasure through forms which, detached from all burdensome meaning, are beautiful, as forms alone'.[23] The film concentrated attention on the juxtaposition of real objects as abstract shapes within frames and emphasizing tempo and movement between them. In such a self-consciously artistic undertaking, critics defended Ruttmann's prerogative to create such a film. On the other hand Paul Friedländer and other contemporary critics also recognized a darker side to this project. As many critics – starting with Carl Mayer himself – have pointed out, Ruttmann's determining influence as editor and director produced a film about Berlin that was superficial. It never got beneath a 'surface approach'.[24] In 1928, Siegfried Kracauer epitomized this perspective:

> A film without a real plot, it attempts to allow the metropolis to arise out of a sequence of microscopic individual traits. But does it convey the reality of Berlin? No: it is just as blind to reality as any other feature film, and this is due to its lack of a political stance. Instead of penetrating its enormous object in a way that would betray a true understanding of its social, economic, and political structure, and instead of observing it with human concern or even tackling it from a particular vantage point in order to resolutely take it apart, Ruttmann leaves the thousands of details unconnected, one next to the other, inserting at most some arbitrarily conceived transitions

> that are meaningless. At best, the film is based on the idea that Berlin is the city of speed and of work – a formal idea that in no way leads to any content and that perhaps for this reason intoxicates the German petit bourgeois when it appears in society and literature. There is nothing to see in this symphony, because it has not exposed a single meaningful relationship.[25]

Though he used 'real' photographic material, Ruttmann's avant-garde editing produced a film that reinscribed and gave voice through its representation of the city and its inhabitants to conservative impulses, which were increasingly prevalent in Ruttmann's Germany. Against expectations given the politically liberal, or even revolutionary stances of Meyer, Meisel and Ruttmann, the film didn't challenge realities. It was not the forerunner of the social documentary form. In its portrayal of the city, the film abdicated its progressive ideological responsibility to uncover hardship and injustice. The combination of Ruttmann's images with Meisel's music might be interpreted as producing a similarly ambivalent critical response. Meisel's music strikingly reinforces the first visual strategy through its indexical, metaphoric and synchronization practice. Called an example of 'extrinsic modernity' by a contemporary music critic, Meisel's musical score was seen as embracing a superficial portrayal of the mechanical and extra-human aspects of modernity.[26] This critical perspective considered the music merely illustrative, though powerful. The music emphasized the train as spectacle. Rather than explore issues such as the social implications of technology, the symbiosis of rural and urban, or the living conditions of those residing by the tracks, this combination of music and image told the story of the train itself. The train symbolized an unstoppable force, which led inextricably to the city, Berlin. In doing so, the film music with the image track reified existing political conditions.

Reverberations of *Berlin* in hindsight, 1985

In retrospect between its première and the 1990s, *Berlin*'s significance lies in two areas: as an influential model on the genre and as a rich trove of documentary source material. Its impact on the international city symphony genre was immediate.[27] City symphony films after *Berlin* might be seen as a concerted effort to produce works that were similarly compelling as art, but with more explicit ethical and political content.[28] The form continued, updated by integrated soundtracks and *cinema vérité* with films such as Chris Marker's *Le Joli Mai* (1963) and Godfrey

Reggio's *Koyaanisqatsi* (1982). But Ruttmann's *Berlin* was largely unseen, at least by American audiences, from the end of the war until the mid-1970s when it again became available for distribution.[29] '*Berlin* is an excellent and always unexpected ingredient in introductory classes', opened Donald M. Bahr's 1979 review of *Berlin* for *American Anthropologist*.[30] Bahr noted that the film was available from The Museum of Modern Art in 'black and white, silent'. The film returned to circulation, and was often used didactically for – and until recent reissues, shown silently to – undergraduates studying Weimar Berlin in the stabilized period.[31] Finally, as a singular source of photographic evidence of a city largely destroyed by the Second World War, excerpts from the film were also used in historical television programmes and documentaries.[32]

In Richard Leacock and Susan Steinberg Woll's 1985 documentary *Lulu in Berlin*, silent film actress Louise Brooks recounts that she rejected a lucrative Hollywood offer in favour of G. W. Pabst's invitation to go to Berlin. Mid-story, Brooks says, 'thank you, Mr. Schulberg. I'll quit and go to Germany', and with that last word the rushing train of Ruttmann's *Berlin* is conjured in the image track. Here it functions as B-roll imagery, explicating the lure of that city in 1928 to an American. In support of that desire, the musical underscoring is a contemporaneous recording of 'The Mooche' by Duke Ellington and his Cotton Club Orchestra.[33] A sinuously improvising clarinet is accompanied by the chugging of a rhythm section. The ensemble functions as a marker of vernacular modernism, up-to-date in its period reference but a product of the international mass culture that this film's original soundtrack represented only obliquely. Furthermore, the music proceeds impervious to the demands of the image, with no important points of synchronization.

Critical perspectives on the film hardened after the war. Many continued to assert Ruttmann's artistic licence to create an abstract work. *Berlin* was simply a continuation of his previous painterly experiments on film. In this view, the film 'merely [took] its raw material from images of the city of Berlin', but as an abstract work of art should not be held accountable for its content.[34] Referencing Bill Nichols's documentary modes, Jill Nelmes asserted that 'the so-called "city symphony" films, such as Walter Ruttmann's *Berlin* [...] may all be seen as "poetic" documentaries – they evoke a mood rather than stating or asserting things directly'.[35] However, contrarian positions continued to be voiced as well. From the vantage point of twenty years later, Kracauer revisited his critique. In *From Caligari to Hitler* (1947), he reasserted that, in *Berlin*, 'Ruttmann, on his part [as opposed to Vertov], focuses upon a society which has managed to evade revolution and now, under the stabilized Republic, is nothing but an unsubstantial conglomeration of parties and ideals.'[36] He summarized this critique by noting that, to his detriment, Ruttmann adopted Russian montage-style editing, but abandoned the revolutionary impulse of the 'Kino-eye' group, which sought to decipher the world through the camera lens with communism.[37] In 1984, Barry Fulks went further. 'Ruttmann's fetishization of the rhythmic and visual as ends in themselves, fused with the cult of technology and urban modernity that characterized the New Objectivity (*Neue Sachlichkeit*), took on aspects of an omnivorous cinematic hubris seeking gratification by the manipulation of what Ruttmann termed the "living material" of a metropolis and the "absolute, purely filmic visual motifs" it yielded.'[38] Fulks carried these implications even further when he observed that there existed a close affinity between National Socialism and the so-called New Objectivity of the 1920s. Given Ruttmann's adaptation of his early cinematic style for Nazi industrial and educational films (*Kulturfilm*) after 1933, it seemed impossible not to question the ideological work that seemingly started earlier with his 'city symphony'.

Soundtracks of *Berlin* through the 1980s do not reinforce either visual strategy: neither spectacle nor continuity editing. The sounds do not operate indexically or present any moments of concordance between sound and image tracks. Screened silently, *Berlin* presented to audiences an audio-visual experience in which the powerful forces behind these images were literally muted. But in a metaphorical sense, silent screenings of the film might be considered to have symbolically 'silenced' critical reception of the film, refusing to engage the implications of the film's social, political and economic positions. By presenting the film silently and considering it as a purely visual work, an art-for-art's-sake consideration of the film would suppress the power relations supported by the soundtrack. With a non-integrated soundtrack such as Ellington's 'The Mooche', the film lends the historical significance of its image track to the music's project. American jazz here metaphorically represents Brooks's individual spiritedness and her autonomy from the Hollywood film industry that sought to harness her youthful energy and sex appeal. The music brings independence and its own narrative agenda to this setting, as uninhibited and unassimilable as Louise Brooks was to Pabst.

Conclusion: Ruttmann's *Berlin* in our times, 2009

More recently, the film has attracted a stream of new musical scores, including a 1986 commission for the Catalan progressive rock band,

Pegasus, from the Barcelona Film Festival; Timothy Brock's lyrical orchestral score (1994) for David Shepard's 1999 restoration, an anonymous electro-acoustic score for a 2002 Divisa release on DVD; and DJ Spooky aka That Subliminal Kid's live accompaniment for a screening at the Tate Modern in 2006. Two reconstructions of Meisel's score were presented live in 2007. One, scored for two pianos with percussion, was a collaboration between the Foundation Deutsche Kinemathek Berlin with the Robert Schumann Institute in Düsseldorf. It was screened with live music at Carnegie Hall as part of the 'Berlin in Lights' festival. The second, a restored version of the film from German television (ZDF/ARTE) and the Munich Filmmuseum, featured a recreation of Meisel's orchestration by Bernd Thewes. It was premièred at Berlin's Friedrichstadtpalast by the Rundfunk-Sinfonieorchester Berlin, and has been seen and heard on broadcasts and on a recently released DVD. The Berlin-based electronic duo Tronthaim have performed their new music to the film at numerous European cultural festivals, including 'Notte d'Estate' in Florence and at the 'Salon du livre' in Paris. Spanish composer Alberto Novoa Rodriguez created and recorded an electro-symphonic soundtrack for this film in 2009. Also in 2009, I composed music for string quartet for the first *Akt*. The recent history of this film's recirculation is equally about its musical settings, which through the addition of different musical scores and new musical means such as electronics and electro-acoustic elements, have assumed quite varied meanings. The film with its new scores is being screened in a multitude of venues such as concert halls, festivals, classrooms, art galleries, broadcasts, fixed media such as DVD, and streamed on the Internet, each with implications for how the film is seen and heard. Each soundtrack presents Ruttmann's city of Berlin anew, sometimes critically, sometimes celebratory. With each new score, the composer asks audiences – sometimes tacitly, sometimes explicitly – to engage in the debate over the combined meanings generated by image and music.

My score privileges the narrative impulses of this sequence, the second of the two visual strategies, by downplaying the mimetic aspects of the music, working against Meisel's metaphoric uses, and drawing the viewer's attention to select concordances between music and image that emphasize continuity editing over the train's visual spectacle. This music is most often performed in a concert setting, an environment of enforced contemplation, by string quartet, which has since the Classical period been associated with independence of musical voice, and the cooperative musical metaphor of 'conversation'. The musical lines are contrapuntal, long and lyrical, performed by solo violin and viola interlocked in hocket with harmonic support in a rocking rhythm provided by the two remaining voices. The phrasing is regular but not entirely predictable; every once in a while things don't quite line up as expected. Voicings emphasize the interval of a perfect fifth, a subtle reference to the train's whistle. Concordances, here in the form of harmonic change, synchronize with select moments of visual interest and emphasize cuts between middle and long shots. For example, the first introduction of the supporting harmonic chord is aligned with the appearance of workers' housing, the first indication of human habitation seen from the train. The harmonic support shifts dramatically again when larger housing blocks are introduced, a symbolic association of the living spaces of unseen people in this sequence with multiple musicians playing together. The entrance to the train station is marked, not with a tempo change, but by having the entire musical ensemble 'hit' the visual triplet of tracks–signal–tracks with a unison rhythmic figure. This soundtrack combines with the image track to privilege the second strategy, and to encourage viewers to identify with the subject position of a passenger on the train. In this way, viewers experience this journey as if the train's commanding power, energy and presence is harnessed for their own ends.

This is not to say that the problematic themes of mass culture presented by this film – surveillance, domination, negative aspects of technology, alienation – are not still easily read from the image track. These are terms very much under negotiation in our own relationship with urban modernity, and a plethora of contemporary scholarly work on this film continues to explore these themes. Recent music has similarly participated in this ongoing assessment of the film, and offers musicians a means to participate in this debate in progress. My soundtrack strives to balance the city's excitement with an acknowledgement of its human participants. It encourages a more sympathetic and humanistic understanding of the individual's relationship to the film's urban environment, and thus offers an implicit critique of mass culture. The film editor Paul Falkenberg, in explaining sound montage and defending Ruttmann said, 'Now, each and every combination of images and sounds produces a strong emotional–psychological effect. The slightest deviation from the original concept may change the meaning of the sequence into its opposite.'[39] Falkenberg meant this in a cautionary sense arguing that we should understand Ruttmann's skilful filmmaking as intentional. Undoubtedly, he's right. The effect of setting different music to an unvarying image track inevitably changes the meanings for viewers – or, in the language of semiotics, such a commutation test shows that these musical signs can powerfully alter the informational content. However, both Ruttmann in his writings and the critics of his Berlin film have emphasized the impact of the visual track to the exclusion of the audible. To the contrary, these examples show that music facilitates different types of viewer engagement with the narrative and formalistic dimensions of the image track, and in doing so, encourages different understandings of the film. For better or worse, the film's meaning is necessarily the result of the combination of image and music tracks. Rather than think of each new score as displacing old meanings, this film has acquired since its creation an accretion of different meanings. The film's original architects strove to create a new form of representation. Students have struggled to imagine the excitement of the stabilized Weimar period. Theorists have engaged issues of critical understanding. Musicians have aspired to participate in the unfinished humanist project of the Weimar Republic and to bring these concerns to the present. For Michel Chion, the visual and auditory in cinema influence each other, and every act of perception necessarily involves a complicated negotiation between the two.[40] With each new musical setting for *Berlin*, the audio-visual contract is subject to renegotiation.

Notes

1. Béla Balázs, *Theory of the Film: Character and Growth of a New Art*, trans. Edith Bone (London: Dennis Dobson, 1952), 133.
2. John Grierson, 'First Principles of Documentary', in *Nonfiction Film: Theory and Criticism*, ed. Richard Meran Barsam (Dutton: New York, 1976), 24.
3. Michael Cowan, *Walter Ruttmann and the Cinema of Multiplicity: Avant-Garde – Advertising – Modernity* (Amsterdam: Amsterdam University Press, 2014), 60–81.
4. Sabine Hake, 'Urban Spectacle in Walter Ruttmann's Berlin Symphony of the Big City', in *Dancing on the Volcano: Essays on the Culture of the Weimar Republic*, ed. Thomas W. Kniesche and Stephen Brockmann (Columbia: Camden House, 1994), 127.
5. Balázs,*Theory of the Film*, 133.
6. Fiona Ford, 'The Film Music of Edmund Meisel (1894–1930)' (PhD dissertation, University of Nottingham, 2011), 149.
7. 'Camera Man Who Made *Last Laugh* Produces Film of a Day in Berlin', *New York Times*, 6 May 1928, 114.
8. My music has been released on CD. Matthew Malsky, *Berlin, Symphony of a Great City Akt I*. QX String Quartet. On *Music for String Quartet* (Baton Rouge, LA: Centaur, 2010). Additionally, it is available with the image track at http://www.clarku.edu/faculty/mmalsky/composer/berlin.html, accessed 1 March 2015.
9. Anton Kaes, 'Leaving Home: Film, Migration, and the Urban Experience', *New German Critique* 74 (1998), 180.

10. Matthew Bernstein, 'Visual Style and Spatial Articulations in *Berlin, Symphony of a City* (1927)', *Journal of Film and Video*, vol. 36, no. 4 (1984), 5.
11. For my analysis, I have relied on multiple sources of the film and Meisel's music. I have consulted two written scores. Edmund Meisel, *Berlin – Die Sinfonie der Großstadt* (Berlin: Ries & Erler, 2011) is a commercially available orchestral score reconstructed from an archival source; the Arthur Kleiner Collection of Silent Movie Music in the Special Collections, Rare Books & Manuscripts division of the University of Minnesota Library contains a complete, printed piano score for *Berlin: Die Sinfonie der Grosstadt* (Box 33, File Folder B33FL). I wish to thank the staff of this facility for their gracious assistance. There is a high degree of corroboration between these sources. I have consulted a commercially available recording of Meisel's reconstructed orchestra score. *Berlin, die Sinfonie der Großstadt*, DVD, directed by Walter Ruttmann (1927, Berlin, Germany: Edition Filmmuseum 39, 2008).
12. Edmund Meisel as translated in Ford, 'The Film Music of Edmund Meisel', 144.
13. Ford, 'The Film Music of Edmund Meisel', 35.
14. Ibid., 39–40.
15. Nora Alter, '*Berlin, Symphony of a Great City* (1927): City, Image, Sound', in *Weimar Cinema: An Essential Guide to Classic Films of the Era*, ed. Noah Isenberg (New York: Columbia University Press, 2009), 196.
16. Ford, 'The Film Music of Edmund Meisel', 147.
17. Ibid., 156–7.
18. Lucy Fischer, '*La Roué*: Movies, Modernity, Machine, Mind', *Modernism/Modernity*, vol. 20, no. 2 (2013), 189.
19. R. Murray Schafer, *The Tuning of the World* (New York: Alfred Knopf, 1977), 230–3.
20. See Matthew Malsky, 'Sounds of the City: Alfred Newman's "Street Scene" and Urban Modernity', in *Lowering the Boom: New Essays on the History, Theory and Practice of Film Sound*, ed. Jay Beck and Anthony Grajeda (Urbana: University of Illinois Press, 2008) for a discussion of this formal plan in the music of Alfred Newman. For a discussion of this 'structural approach' in the music of Antheil for *Ballet mécanique* (1924), Satie in *Entr'acte* (1924), and Meisel for *Bronenosets Potyomkin* (1925) and *Oktabr* (1927), that emphasizes music's temporal rather emotional content and effects, see Julie Hubbert, 'Eisenstein's Theory of Film Music Revisited', in *Composing for the Screen in Germany and the USSR*, ed. Robynn J. Stilwell and Phil Powrie (Bloomington: Indiana University Press, 2008), 127–36.
21. Ford, 'The Film Music of Edmund Meisel', 145. Also for further subsequent uses of this device by composers Hindemith and Eisler with abstract films by Ruttmann and Richter at the Baden-Baden music festival, see Malte Hagener, *Moving Forward, Looking Backward: The European Avant-Garde and the Invention of Film Culture 1919–1939* (Amsterdam: Amsterdam University Press, 2007), 141–3.
22. Syncresis is a portmanteau of Michel Chion (of synchronism + synthesis). It is a psychological phenomenon that makes post-production synchronized sound techniques. For example, there are many possible voices that will sound convincing when dubbing (i.e. multiple voices are believable when

combined with the same filmed person). Michel Chion, *Audio-Vision: Sound on Screen* (New York: Columbia University Press, 1994), 63–4, 224.

23. Barry Fulks, 'Walter Ruttman, the Avant-Garde Film and Nazi Modernism', *Film & History,* vol. 14, no. 12 (1984), 28.
24. Siegfried Kracauer, *From Caligari to Hitler: A Psychological History of the German Film* (Princeton, NJ: Princeton University Press, 1966 [orig. 1947]), 184.
25. Siegfried Kracauer, 'Film 1928', in *The Mass Ornament: Weimar Essays,* trans. and ed. Thomas Y. Levin (Cambridge, MA: Harvard University Press, 1995), 318.
26. Ford, 'The Film Music of Edmund Meisel', 151.
27. Alter, '*Berlin, Symphony of a Great City* (1927)', 196–7.
28. Examples of city symphony films following *Berlin* include Dziga Vertov's *Man with a Movie Camera* (Soviet Union, 1929), Jan Vigo's *À propos de Nice* (France, 1929), Robert Florey's *Skyscraper Symphony* (USA, 1929), Herman Weinberg's *A City Symphony* (USA, 1929), Kenji Mizoguchi's *Takahiko kyogoku* (Japan, 1929), and Adalberto Kemeny and Rudolf Rex Lustig's *São Paulo, sinfonia da metrópole* (Brazil, 1929).
29. Mention of Ruttmann's film in the *New York Times,* while not an exhaustive form of evidence, is evocative. There were only two screenings listed between 1930 and 1969. *Berlin* was shown at the Museum of Modern Art on 28 May 1939, as part of a public series of 'famous films' from their film library. A second screening is listed at the Fifty-Fifth Street Playhouse on 4 December 1954.
30. Donald M. Bahr, 'Berlin', *American Anthropologist,* vol. 81, no. 1 (March 1979), 209.
31. Personal correspondence with Germanists Alan Lareau and Robert Deam Tobin.
32. Kay Hoffmann, '*Berlin: The Symphony of a Great City* (Germany, Ruttmann, 1927)', in *Encyclopedia of the Documentary Film,* ed. Ian Aitken (New York: Routledge, 2006), 110–11.
33. Duke Ellington and his Cotton Club Orchestra, 'The Mooche', *The OKeh Ellington.* Columbia CK 46178, 1991/1927, compact disc.
34. Walter Schobert, '"Painting in Time" and "Visual Music": On German Avant-Garde Films of the 1920s', in *Expressionist Film: New Perspectives,* ed. Dietrich Scheunemann (Rochester, NY: Camden House, 2003), 242–3.
35. Jill Nelmes (ed.), *Introduction to Film Studies* (New York: Routledge, 1996), 212.
36. Kracauer, *From Caligari to Hitler,* 185–6.
37. Ibid., 185.
38. Fulks, 'Walter Ruttman, the Avant-Garde Film and Nazi Modernism', 28–9.
39. Paul Falkenberg, 'Sound Montage: A Propos de Ruttmann', *Film Culture,* no. 22–3 (1961), 59.
40. Michel Chion, *Audio-Vision: Sound on Screen* (New York: Columbia University Press, 1994).

Bibliography

Alter, Nora, '*Berlin, Symphony of a Great City* (1927): City, Image, Sound', in *Weimar Cinema: An Essential Guide to Classic Films of the Era,* ed. Noah Eisenberg (New York: Columbia University Press, 2009), 193–216.

Anon., 'Camera Man Who Made *Last Laugh* Produces Film of a Day in Berlin', *New York Times*, 6 May 1928, 114.

Bahr, Donald M., 'Berlin', *American Anthropologist*, vol. 81, no. 1 (March 1979), 209.

Balázs, Béla, *Theory of the Film: Character and Growth of a New Art*, trans. Edith Bone (London: Dennis Dobson, 1952).

Bernstein, Matthew, 'Visual Style and Spatial Articulations in *Berlin, Symphony of a City* (1927)', *Journal of Film and Video*, vol. 36, no. 4 (1984), 5–12.

Blakeston, Oswell, 'Interview with Carl (Karl) Freund', *Close-Up* (January 1929).

Chion, Michel, *Audio-Vision: Sound on Screen* (New York: Columbia University Press, 1994).

Cowan, Michael, *Walter Ruttmann and the Cinema of Multiplicity: Avant-Garde – Advertising – Modernity* (Amsterdam: Amsterdam University Press, 2014).

Ellington, Duke and his Cotton Club Orchestra, 'The Mooche', *The OKeh Ellington*. Columbia CK 46178, 1991/1927, compact disc.

Falkenberg, Paul, 'Sound Montage: A Propos de Ruttmann', *Film Culture*, no. 22–3 (1961), 59–62.

Fischer, Lucy, '*La Roué*: Movies, Modernity, Machine, Mind', *Modernism/Modernity*, vol. 20, no. 2 (2013), 189–211.

Ford, Fiona, 'The Film Music of Edmund Meisel (1894–1930)' (University of Nottingham, PhD thesis, 2011).

Friedlander, Paul, 'Berlin: die Symphonie der Grossstadt', *Die Rote Fahne*, 25 September 1927.

Fulks, Barry, 'Walter Ruttmann, the Avant-Garde Film and Nazi Modernism', *Film & History*, vol. 14, no. 12 (1984), 26–35.

Grierson, John, 'First Principles of Documentary', in *Nonfiction Film: Theory and Criticism*, ed. Richard Meran Barsam (New York: Dutton, 1976), 19–30.

Hake, Sabine, 'Urban Spectacle in Walter Ruttmann's *Berlin Symphony of the Big City*', in *Dancing on the Volcano: Essays on the Culture of the Weimar Republic*, ed. Thomas W. Kniesche and Stephen Brockmann (Columbia: Camden House, 1994), 127–42.

Hoffmann, Kay, '*Berlin: The Symphony of a Great City* (Germany, Ruttmann, 1927)', in *Encyclopedia of the Documentary Film*, ed. Ian Aitken (New York: Routledge, 2006), 109–11.

Hubbert, Julie, 'Eisenstein's Theory of Film Music Revisited', in *Composing for the Screen in Germany and the USSR*, ed. Robynn J. Stilwell and Phil Powrie (Bloomington: Indiana University Press, 2008), 125–47.

Kaes, Anton, 'Leaving Home: Film, Migration, and the Urban Experience', *New German Critique*, no. 74 (1998), 179–92.

King, Norman, 'The Sound of Silents', in *Silent Film*, ed. Richard Abel (New Brunswick, NJ: Rutgers University Press, 1996), 31–44.

Kracauer, Siegfried, 'Film 1928', in *The Mass Ornament: Weimar Essays*, ed. and trans. Thomas Y. Levin (Cambridge, MA: Harvard University Press, 1995), 307–22.

Kracauer, Siegfried, *From Caligari to Hitler: A Psychological History of the German Film* (Princeton, NJ: Princeton University Press, 1966 [orig. 1947]).

Malsky, Matthew, 'Sounds of the City: Alfred Newman's "Street Scene" and Urban Modernity', in *Lowering the Boom: New Essays on the History, Theory and Practice of Film Sound*, ed. Jay Beck and Anthony Grajeda (Urbana: University of Illinois Press, 2008), 105–22.

Meisel, Edmund, 'Wie schreibt man Filmmusik', *Ufa-Magazin*, 1–7 April 1927. Reprinted in *Der Stummfilmmusiker Edmund Meisel*, ed. Werner Sudendorf (Frankfurt am Main: Deutsches Filmmuseum Frankfurt, 1983), 57.

Nelmes, Jill (ed.), *Introduction to Film Studies* (New York: Routledge, 1996).

Schafer, R. Murray, *The Tuning of the World* (New York: Alfred Knopf, 1977).

Schobert, Walter, '"Painting in Time" and "Visual Music": On German Avant-Garde Films of the 1920s', in *Expressionist Film: New Perspectives*, ed. Dietrich Scheunemann (Rochester, NY: Camden House, 2003), 237–49.

Spratt, Geoffrey K., *The Music of Arthur Honegger* (Cork: Cork University Press, 1987).

12
Silent Film, Live Music and Contemporary Composition

Ed Hughes

Silent film and live music together force you to think differently about both phenomena. And so, although composers frequently engage with external elements such as texts and non-musical subjects, it is my contention that working with silent film as a musician raises especially interesting challenges. In this chapter, I outline my previous composition work with silent moving images, and conclude with a case study of a recent collaborative project, *Night Music* (2015). One thing my examples have in common is that they were all conceived for live musical performance with silent film screenings – although some projects have since been released commercially on DVD. This latter point speaks to some recent film music studies research which reminds us of the affective loss sustained in the transition from silent to sound film in the late 1920s and early 1930s: silent film music practice rests on a visible distinction between the screen and the source of sound and music, as in opera where the numinous and magical quality of the action is possible precisely because of the separation between stage and audience by the orchestra. I argue that my own live music and screen work fits in with a tradition of seeking to recover this quality through performances featuring live solo musicians and orchestras, and furthermore that the ritual, simultaneous working through of screen and musical ideas that this work displays aligns it more with certain qualities of opera than with conventional film music practices found in sound cinema.

My usual practice as a composer begins with an opportunity or idea that is founded in some practical challenge or problem, but typically not fully formed (at its simplest – a musician would like a new piece to perform in a concert). The next stage in most such projects is to find some focus. For me – in the absence of extra musical stimuli – this typically involves reflection, sketching, re-sketching and further reflection on musical material, until a forward compositional momentum takes over. Musical material means ideas formed from elements such as pulse, tempo, rhythm and in my case mostly adherence to the Western tempered pitch system. Working with these elements I find myself returning always to fundamental concerns with musical intervals, harmonies, pitch centres and lines. These concerns inform everything I write. There is a background preoccupation with certain intervals and harmonies which seems hard-wired. Within this framing, I then find flexibility and freedom to write, which occasionally leaps into a sense of fluency at certain points in the writing process, which you recognize when you're there but which is very hard to analyse or describe. As you write, what you are trying to write becomes gradually clearer and reveals itself, if you are lucky, to the point where external links, ideas and comparisons may become viable.

For example, in 2010 I was invited to contribute to *Signals*, an album of new solo oboe works.[1] The brief was: a composition of four to five minutes, for solo oboe – apart from this, completely open. The composition began with improvisation of a musical idea – a single recurring interval – minor seventh. The repetition of this interval pattern establishes a sense of harmony and pitch centre. The music acquires tension as it shifts away from this opening and explores wider registers. I composed this piece following purely musical lines of thought; yet in the background were very real-world thoughts because at the time I was also working on another composition that was very political: a setting of Guantánamo Bay internees' poems to protest against detention without trial. An important theme in some of these poems was separation and distance from beloved families and home. As I worked on the solo oboe piece I realized that the abstract musical tension between the initial musical idea with its emphasis on E and the minor 7th interval, and the distant and remote places the oboe travelled to in the course of the piece, could embody this theme of distance and separation, expressed powerfully in the poems. And so I entitled the solo oboe piece 'As Dreams Begin' after a poem by Moazzam Begg, a former internee.

I mention the solo oboe example because it is an example of practice that was, *at the time of composition*, not consciously affected by external elements, such as narrative and images. The piece attracted itself, as it were, to an external poem, because an alignment in poetics arose between the intrinsically musical qualities of the piece and the images of the text. However, in the last decade or so, alongside composing concert music, I have been drawn to working directly with film, and have here found a productive tension between the abstract and autonomous nature of musical composition, and the requirement to respond to external narratives and technologies, very exciting and challenging to my techniques and methods as a composer.

For example, for the 2005 Brighton Festival I was commissioned to write a new score to Eisenstein's film *Battleship Potemkin* (1925), on the occasion of the film's eightieth anniversary (Figure 12.1). The opportunity was stimulating both in terms of the scale of the work – I had not previously written a continuous musical work of 74 minutes in five episodes – nor had I before responded to a radical visual language. I have since often thought that the experience had a decisive effect in influencing the extension of my later concert and operatic music. I chose to write my *Potemkin* score for a tight ensemble with orchestral and percussive qualities and an emphasis on granular and earthy timbres (flute, clarinet, horn, trumpet, piano, percussion, cello and double bass). These qualities were extended through the use of four channel electronics, which spatialized the sound with special attention to the percussion part at a critical point in the film – the famous Odessa Steps scene. This latter scene is compelling to watch and demanding to compose because, although one recalls the famous stills of the woman with smashed glasses (which inspired Francis Bacon) and the pram descending the steps, it is easy to forget that these fragmentary images appear fleetingly in a relentlessly sustained visual line of 7 minutes 25 seconds, or nearly 10 per cent of the running time of the film, from the first time the Cossacks fire on the innocent bystanders all the way to the symbolic destruction of the opera house and the roaring stone lion. Eisenstein, reflecting on his supervision of the German composer Edmund Meisel in his score for the Berlin Film Festival performance in 1926, commented 'even now sound film music is almost always "close by the film" and doesn't basically differ from former "musical illustrations". However [I was in Berlin long enough] to make arrangement with the composer Meisel about the decisive "effect" of the music for Potemkin.' Reading this account guided me to compose music that matched the visual rhythm of the picture and to respond musically to Eisenstein's declared ambition to create pictures capable of 'a qualitative leap', exploding stylistically beyond the limits of the system of 'silent picture with musical illustration' into a new area: 'a unity of fused musical and visual images'.[2]

I followed *Battleship Potemkin* with a further score (2007) to Eisenstein's first full length silent film, *Strike!* (1924). I used identical forces, including electronics, but conceived this as a live accompaniment to screenings across diverse venues including concert halls and

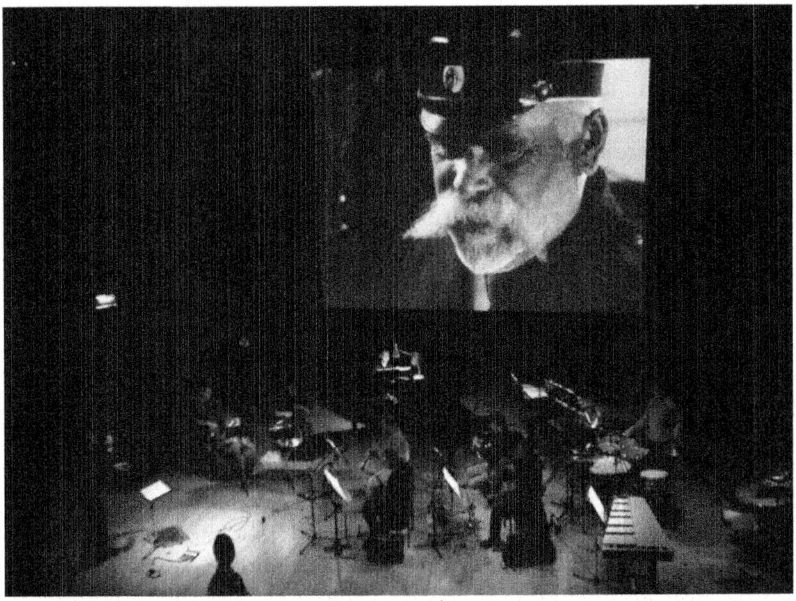

Figure 12.1 Ed Hughes's score to *Battleship Potemkin* in performance by the New Music Players conducted by Patrick Bailey at the Jack Lyons Concert Hall, University of York, 4 November 2005.

the British Library Atrium as part of an Arts Council England supported tour. *Strike!* is visually experimental as well as politically charged – ranging from slapstick to violence to abstract plays of light; it is less unified, perhaps, than *Potemkin*, and pushed me to consider musical form as active, dynamic and changing, rather than as a series of separate vessels to be filled with notes, which was characteristic of the previous work. Although eclectic, each section, or act, of Eisenstein's work precipitates the next; everything is necessary. At the climax of the film, there is a famous scene featuring montage – scenes of slaughter in the abattoir are intercut with the brutal cutting down of the fleeing workers by the Tsarist troops. However, to consider this montage purely in terms of these famous frames is to take them out of context. This extended final sequence stretches back to the furious exchanges between the police and the captured strike leader. As it dawns on the strike leader that his people are at risk, a view of the tenements is briefly cut in. The sequence leads seamlessly into the scenes of mayhem and brutal attack, symbolized by the montage, and the tragic close of the film. My point is that Eisenstein creates breathtaking energy through this sequence using closely constructed rhythmic and polyphonic montage, and not merely a pasting together of non-linear images; there is a sense of line which propels the scenes forward rhythmically but there is also deliberate and designed dissonance in the vertical construction of the images. These extraordinary silent moving images made me reconsider my approach to musical form and musical ensemble writing. In this case, assisted by electronic musician Ed Briggs, I let in previously unconsidered raw sounds of analogue synthesizers to coexist alongside the rhythmic structures of the smooth classical acoustic ensemble – matching Eisenstein's visual line – but providing tension and conflict with the timbres of the acoustic classical ensemble: the rhythmic line in the music propels forward, but the timbres are dissonant and mixed, like the visual constructs in Eisenstein's montage. Thus Eisenstein's *Strike!* invaded my consciousness and spoke of how music could be multiple, splintered and yet energized, and has since opened up new possibilities in my concert music, including my instrumental composition *Chamber Concerto* (2010).

One concern which connects all my music is polyphony – in terms of close note-to-note counterpoint but also through something that places me closer to modernist concerns, namely the counterpoint and elision of larger surfaces and textures. Related to this is my fascination with the work of early English composers such as Thomas Tallis (*c.*1505–85), who in a pretonal era succeeded in harnessing massed voices and triadic harmony to extraordinary spatial effect. The influence of early music, and colliding and interweaving planes of music, comes through in a commission from 2007 for Glyndebourne Opera and Photoworks. The filmmaker Sophy Rickett created a 22-minute film *AUDITORIUM* (2007) which turns what might have been a conventional documentary about the opera house into a silent, abstract, split-screen artistic work in which the human figure is almost entirely absent. I was co-commissioned to write the score for live orchestra and electronics (2007). What was wonderful about Sophy's approach was that it was both silent and musical. Her two screens, constantly up and slowly changing, speak like two voices in a modern composition by the great twentieth-century composer Morton Feldman. Sophy's visuals suggested an approach in which harmonies were to be simplified, and yet the overall compositional effect was made more complex by a layered approach – in the sense of elision and horizontal layering and also spatialization. At the first performance in Glyndebourne Opera House on 17 November 2007, the orchestra was positioned – very visibly – on

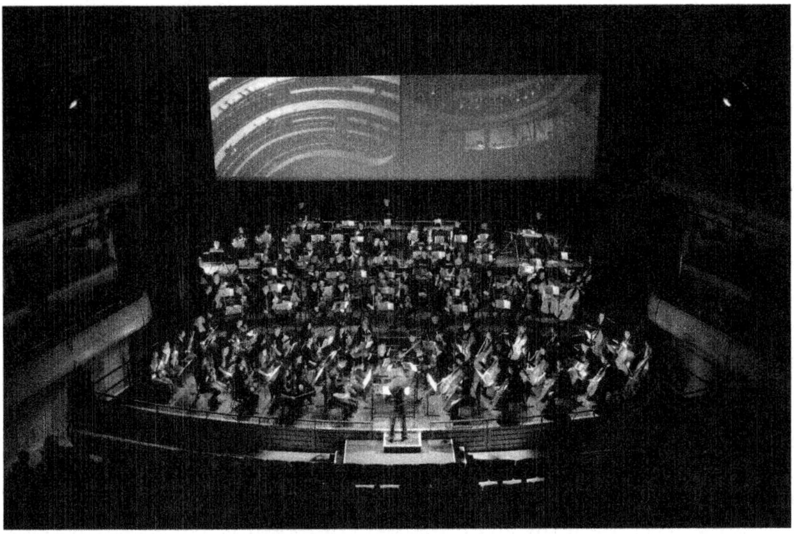

Figure 12.2 First performance of *AUDITORIUM*, a film by Sophy Rickett with music by Ed Hughes, at Glyndebourne Opera House on 17 November 2007. Photograph by Richard Rowlands.

the stage in front of the large screen, and electronic sound was diffused from in front of and behind the audience in the stalls (Figure 12.2). This immersive aspect is retained in the properly installed version of this piece for art galleries, where you hear the score in four channel sound. Memorably, in the De La Warr Pavilion, the film was projected onto Perspex screens hung in the centre of the black space, so that the audience could walk around the image and be surrounded by the sound. For the audience at the live performance première, the experience was of being in the auditorium itself, spectators to an unusual journey through architectural and mechanical space, immersed in sound from the live orchestra and a disembodied orchestra and operatic voices, all of whom are off screen. At the centre of the film, high up in the grid of the house which the audience never sees, a mini opera is enacted using recorded singing voices and multichannel techniques, a musical polyphony created with human voices and technology, corresponding to the lines and virtually abstract rolling shadows in certain sections of the visuals.

It was exciting to work on this film and, in reviewing this work, to be reminded that certain theories of early silent film and music practice hold true for contemporary work in this field. James Buhler and David Neumeyer comment:

> Warner Bros. [...] thought that dialogue and synchronised onscreen musical performances would, like colour, remain a special effect and that synchronised rather than talking film would become the dominant form of sound film. Hindsight allows us to see that this assessment was radically mistaken because it ignored the structural shift in the relations between image and sound. [...] As Paolo Cherchi Usai reminds us, the aesthetic of silent cinema was based on maintaining 'a clear distinction between an apparatus producing images and a sound source in front of or behind the screen'.[3]

The loss of this 'clear distinction' can be seen as part of a wider loss of creative potential in live music and the silent screen in the historical transition of the late 1920s to sound film, effected through the mechanical linking of recorded images and sound. That loss has left a space – meaning the magical separation of the screen from the audience by the presence of an orchestra, ensemble or cinema organ – which filmmakers, musicians and live event producers continue to investigate. I see my work with silent film as part of this movement which is why I emphasize the importance of live music in the projects I do, and also why I increasingly think of live music to picture as a form of 'intervention' rather than accompaniment – because of the special qualities inherent in live performance with screened images.

A new challenge to my composing methods came with an audio-visual project, *Dark Formations* (2010). This project involved working collaboratively with a curator-director in a joint effort to understand and interpret still images presented in silent succession with live musical accompaniment. This project was the idea of art historian Professor David Chandler, who discovered aerial photographs of great power from the Second World War in the archives of the Imperial War Museum, and obtained the support of the late Roger Tolson, the Museum's former Head of Art. These are documentary images of the reality and consequences of Allied bombardment of Germany in 1943–5. Chandler arranged these images into a twelve-minute silent film sequence; the music I created for this for small ensemble was deliberately understated and highly repetitive. This created an effect of recognition and distancing, qualities intensified in the music because it was shorn of expressive features such as string vibrato. In some live performance versions of this I changed the piano timbres even more drastically, by applying ring modulation to the piano. This is an electronic music technique in which the piano's live sound is modulated by a machine-generated waveform; it was used in live performance by Karlheinz Stockhausen and others, in order to modify dramatically the sound of the instrument so closely associated with nineteenth-century German culture, and in particular Beethoven. The refusal of literal sound effects and the presence of strange music is part of a knife-edge balance between documentation and interpretation which makes the implications of these archive images more vivid. As David Chandler writes, 'But in this image, so disturbingly unreal in relation to the circumstances of its particular time and place, we might imagine that the grinding noise of huge machines in space, the shuddering motion of the planes, and all the explosions and blinding flashes had been temporarily suspended as the Lancaster glides silently across the sky casting great arcs of light as if writing its own airborne signature' (2011). For Roger Tolson, the exercise, although conceived for live concert performance and screening, was suggestive in terms of future possibilities for new approaches to installing image sequences with music in museum settings: 'the methodology of *Dark Formations* has been a direct stimulus to how to re-think the selection, narration and display design of disparate groups of material, and one that I believe will enrich these and future displays'.[4]

I returned to this complex and yet all too relevant subject – Europe in flames – in 2014 in response to a request for a new piece from the Brighton-based promoter Music of Our Time. The proposed concert would examine the phenomenon of war and how it is reflected in music. The new piece, *Night Music* (2015), was influenced by *Dark Formations,* in that it also used archive materials from the Imperial War Museum to give a portrait of RAF action over Europe and its effects. The archive silent film materials were prepared in a film devised by Svetlana Palmer and directed by Paul Bernays of Artisan Pictures. Unlike *Dark Formations,* in this case moving images were mined from the archive for the event. The strangeness, proximity and violence of these historical images were magnified by the fact that they are mute. Further, the director of the film, Paul Bernays, intensified their impact by presenting certain passages in slow motion, and adopted a hallucinatory and musically motific approach in the editing.

Night Music is, in its full version, 28 minutes long. It was premièred in the Brighton Science Festival on 12 February 2015 with soloist Joseph Houston, the University of Sussex Symphony Orchestra, Danny Bright (sound diffusion), Paul Bernays (film direction), and Ed Hughes (conductor), for an audience of 185 (including 76 under the age of 26)

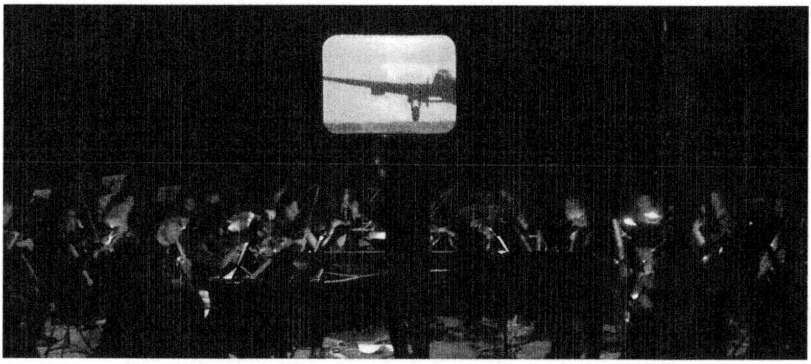

Figure 12.3 Performance of *Night Music* at the Brighton Science Festival, 12 February 2015. Soloist Joseph Houston (piano) with University of Sussex Symphony Orchestra conducted by Ed Hughes.

with 108 orchestra and chorus; thus a combined total of 293 present (Figure 12.3). The work is organized in sections. It presents silent moving images throughout. The audience hears newly recorded commentaries from a veteran member of the crew and a woman who experienced the attack from the ground, alternating with three musical interventions. The structure is in the form of tableaux alternating spoken content with musical commentaries. These tableaux provide the sonic and musical design that gives the basic structure for the work in this form. They, in my view, create a necessary formality which enables the spectator to appreciate unsentimentally the experiences of two individuals caught up in conflict, against the magnitude of the events depicted. This decision, to alternate the audio commentary with musical interventions, rather than to 'duck' the music to overlay audio elements, was one Paul Bernays and I agreed on at our planning meeting on 16 December 2014. It moved the entire project towards a musical footing; the pattern of commentaries alternating with musical interventions recalls patterns in opera in which reflection and action alternate and exchange through the vehicles of arias and ensemble numbers.

This decision to articulate highly representational themes through musical means helped us because we absolutely wished to avoid creating the impression of a conventional film or conventional film music, of 'illustrating' the picture, to quote Eisenstein. In other words we did not wish to glamorize the subject in any way. That said, a moral question persists. Is there really any purpose or useful value in creating art on such a complex and difficult historical subject? Clearly such an undertaking is not an exercise in objective or traditional scholarship. While it was our aim to employ creative musical and media practices, to curate, display and interpret some highly particular historical materials which are rarely seen, this anxiety persisted in my own mind up to the night of the first performance.

A further concern, particular to my position and focus as a composer, was the question of lyricism and harmony, which, though not naively tonal or minimal, my musical language certainly does exhibit. Should one strip out lyricism and harmony? This is not the case in war paintings, in which it is possible to derive, perhaps troublingly, aesthetic pleasure from the presentation of bloody battle scenes. Initial audience responses suggested that there were positive feelings of affect, even though the subject was recognized to be sad or even discomfiting. Further, I would contend again that the music's lyricism can be heard to register human presence and experience of the individuals caught up in conflict. In some senses, this work fits with a long tradition of art which reflects war, and specifically European conflict. As Peter Paret writes:

> Works of art rarely convey reliable information on how large numbers of men in a particular period went about the business of fighting and killing [...] what art conveys best, and sometimes uniquely, has less to do with the mechanics of war – whether these are depicted accurately or in a stylised manner or allegorically – than with the feelings about war of individuals and societies with their attitudes towards the enemy and their own armed forces.[5]

In returning to my main question on the challenge posed by silent film and live music to contemporary composing practices, I will articulate the relationship between the various elements of *Night Music* in order to demonstrate that this project has raised questions about the nature of the musical score in a multi-media setting; *Night Music* was and is a collaborative work, like other compositions with moving images, but with a wider variety of interventions impinging on the three-movement piano score, which is at the heart of the musical response. Unlike the solo oboe work discussed earlier, *Night Music* was conceived, from the outset, in relation to the simultaneous presence of a larger, darkly looming narrative; it was indeed conceived as a counterpoint to moving images, even though the music was written first, and the images were skilfully edited to match the paragraphs of music.

The constituent elements of *Night Music* are:

- pianist interpreting a fixed score;
- strings interpreting unfixed score guided by conductor;
- computer listening and responding guided/shaped by sound designer;
- visuals guided by a film editor who shapes them in real time based on timings derived from rehearsals; also cues to some extent in response to musicians (in the live performance musicians sometimes follow video, and vice versa);
- recorded commentaries in the form of recollections of a surviving RAF crew member, and a German woman survivor of the Hamburg attack.

The *Night Music* piano score is a complex notation of pitches and rhythms directed alternately towards creating flowing surfaces, and, at other times, aggressive, disruptive and even violent events, especially in Parts 2 and 3. I have a strong interest in piano writing. I like the implicit orchestral qualities of the piano, its ability to produce complex chords, lyrical lines and percussive effects and even clusters. My first composition teacher was Michael Finnissy, who created a famously complex, meticulously notated piano cycle, called *English Country Tunes* (1977). *Night Music* is in this tradition: it is fixed, detailed and demanding; yet care was taken to make it legible to its professional performer. But, although as a composer, the score is central to my practice, it is in the performance that the music speaks its true complexity – the score cannot capture, represent or control all nuances of performance. And, arguably, performing relationships become still more complex in multi-media settings.

If the pianist's score is fixed, the string orchestra's is looser – lending itself to a much greater degree of variability from performance to performance. The score consists mostly of a succession of sustained chords, and these are cued by the strings' conductor, largely independently of the solo pianist. Although free, the strings' chords have a definite relationship to the piano music – they reveal the underlying chord sequences that drive the complex polyphony of the piano, but not necessarily synchronously.

The computer is still less fixed and on a note-to-note level would be impossible to transcribe or to notate, although the broad changes to effects required are indicated in the score. I like to think of the computer in this work as listening to the sounds of the piano, and singing along with it, a bit like an unselfconscious child. I use various techniques, but, in a nutshell, the computer (a) listens to and analyses the pitches and matches them with its own sounds; (b) records small snatches (samples) of sound, reverses them and throws them between the loudspeakers to create spatial canons; (c) records sounds and slices them into small fragments to create new textures. These are familiar ideas in live electronic music. But, for me, there is a particular interest in that they add an element of contingency into the mix with the 'fixed' elements of the performance. The counterpoint that is created is interestingly chaotic and yet still tethered to the performance. This is because there is always a harmonic relationship between what the computer is doing and the notes that are being sounded by the piano. So there is a chaotic layer, mitigated by control – the control and shaping of the human performer at the computer, who is participating in and responding to the performance, meaning the gestures of the musicians and the unfolding narrative on the screen.[6]

Thus, there are in *Night Music* at least three unfolding planes of music, moving at different speeds, which sometimes meet (in the form of a beat or accent) but often do not; and similarly these musical weavings periodically find a beat or accent in the screen visuals, forming an anchor point (in an echo of Eisenstein's theory of audio-visual counterpoint). The fact that there is not a simple unitary system, as in my oboe solo, but a polyphony of elements, is bound to make the relationship between the spectator and the archive material more complex. The spectator has to see the visual material through layers of auditory musical material, as it were. To compensate for this deliberate 'difficulty', the film is presented in slow motion during the musical interventions. Again, this speaks to an operatic quality: harnessing the potential of live music and silent film to arrest and to accelerate time. My hunch is that this constellation of elements confers critical distance that a simpler, unitary relationship would not offer: the viewer is moved to a critical position by having to deal with the shocking immediacy of images of burning and destruction at the same time as experiencing a sense of distance, strangeness and remoteness imposed by the filters and layers of the music, and the slow and silent motion of the images.

In this sense, the music returns to the moral question – how can you write music to accompany such images? By asserting that its function is not to illustrate action in the fashion of a Hollywood film but rather to render this originally functional footage complex, and carry or express something of what we might be able to understand in the human experience of war. This is music that has not renounced lyricism; but what is there to sing about? Perhaps, the experience of the individual – the fact that the individuals in war should have voices, including pilots, navigators and victims on the ground. Music and silent film carry many voices, wordlessly. It's not illustration (a normal understanding of film music) – it's a voicing of off-stage presences.

Analysis of film to music relations in *Night Music*

Reflecting on the three movements of *Night Music*, I find that the collaboration has produced an evocation of the experience of war through a balance between the inherent qualities and possibilities of film and of music; and the association with what is represented. Thus, it is possible to perceive the internal patterns, repetitions, variations and structures of music and film anchoring themselves to certain themes that are depicted. Here is my personal and non-prescriptive interpretation of the musical tableaux:

In *Night Music* Part 1 the film shows crew scenes in which comradeship is discernible. But we cannot hear the conversations because the material is silent; the piano, with its subjective and elaborate melodic lines, becomes the voices of the individuals on the ground. The strings are more neutral – their slowly moving chords amplify the harmonies underpinning the piano music but seem to be actually more concerned with rendering context. One might even associate the strings with the earth, unnaturally violated by deadly machines in Parts 2 and 3 of *Night Music*. The computer sounds are contingent on the acoustic musical material, and yet also bear free elements and unexpected artefacts within an overall envelope shaped by the human computer sound controller. The computer sounds thus encapsulate something of the sense of absolute determinism inscribed into the mission with the aspects of extreme chance that affect its outcome. The computer sounds, in their uncontrolled aspects, could be associated with the sky, and flight.

In *Night Music* Part 2 the musical materials are clearly concerned with gestures of ascent and descent, corresponding to the mechanical effort involved in getting the plane into the air. In fact the music continually balances mechanistic with lyrical qualities, because there is a tension between the machinery and the strange, remote poetry of flight. Recalling the title of my previous collaboration, with David Chandler, we see the aircraft moving as a group, in dark formation over dazzling white clouds. The middle section of the work consists of punchy, short, sharp chords which are echoed in the *sforzando* attacks of the strings group – perhaps corresponding to the violation of the ground from the air.

In *Night Music* Part 3 we begin visually with inky blackness punctuated by intense white, pulsating flashes, which are images of Hamburg aflame from the air. These are echoed, not synchronously but later in the piece, by the most violent episodes in the music, comprising virtuoso keyboard writing cut through with harsh chords and clusters. At the start of *Night Music* Part 3, musically, there is a flowing toccata in marked contrast to the slow motion of the film; the fractured harmonies that impinge on this flowing surface suggest the inevitability of conflict. The very poignant cut to the ground scenes in the aftermath of the attack is accompanied by music of objectivity (the film director Bernays spares us very difficult footage at this point but the scenes are nevertheless shocking, perhaps because also contemporary in their resonances). The big cluster chords' sequence in the piano, mentioned earlier, is matched to the fragility of ground scenes and darkness. The close of the piece musically recalls the tonal area of the opening, as though a shadow of a nineteenth-century musical form; an echo of the structure of a Beethoven sonata. The visual repetition by Bernays of the single, burnt, isolated standing figure is a poignant motif which balances this musical strategy.

Findings

As a follow-up to the *Night Music* project I surveyed a small sample of the audience with a brief questionnaire via a Google Docs form, which was voluntary and anonymous. The comments appeared to confirm that the music in conjunction with the film created impressions and an experience that would not have arisen if the audience had only seen this footage without music or bespoke editing.

> *A1. The music especially seemed to fit the film and accentuated the atmosphere of the whole piece. The music was difficult to listen to at times in terms of the discordance but overall I feel that this added more to the mood of the performance.*

> *A3. It would have been a completely different experience without the music. It both enhanced and complimented the film, which created a moving and powerful experience.*

> *A4. I thought it was an interesting compositional decision to understate the musical element and no attempt made to create an atmosphere of terror. A good choice to my mind.*
>
> *A7. It created an immersive experience. The footage was moving and the music enhanced the emotional charge of the film.*

The intervention of the music, and the slowing of the archive material to a non-naturalistic pace, seemed to open out the possibility of a different kind of counterpoint and mode of appreciation, akin to reflective points in an opera. I have described these moments as 'tableaux'.

> *A5. A cohesive, unusual construction. Image editing options seemed quite idiosyncratic, which was probably helpful given that archive images can trigger a sense of recognition whether or not we've actually seen them. This added to the sense of a wilfully poetic work, rather than of a 'documentary with original musical accompaniment'.*

Another member of the audience commented: *I could feel a tussle between the beauty of those aerial view shots and the horrific reality of wartime. So emotive cutting directly between hearing both verbal and musical accounts/reflections of war* (Laura Callaghan Groomes – emailed comment 13 February 2015).

Knowledge of the history of Bomber Command was affected by seeing the documentary elements mixed with slow motion archive film and music. For many people such material is familiar, but for some it is not.

> *A3. I already knew some history of the Bombers, but had not ever seen film images of the actual bombing of Germany, taken from the planes. It was really interesting, but quite harrowing as well.*

The effect of 'working through' these images, repetitively, slowly, non-literally and with music, recalls an idea of John Ellis who, in connection with television, writes that the repetition and rehearsal of images is a crucial therapeutic activity in an open and 'information-rich' environment.[7] Situating this material in the artistic context of a concert with live music and screen – and then again within an overall form that, in certain aspects, is analogous to the action of opera, brings a new spin to the concept of 'working through' both familiar and unfamiliar images. The direct treatment of the images in slow motion, and their combination with silence and cyclical musical structures, enables one to scrutinize the material anew in ways that seem to release a positive affect. For one spectator (A3): *it would have been a completely different experience without the music. It both enhanced and complimented the film, which created a moving and powerful experience.* Thus through a creative intervention involving non-literal and quasi-operatic orderings of live music and silent film, the images and themes of an important and traumatic event of July 1943 were relived and worked through by a contemporary audience.

The experience overall made me more conscious than before that external elements (histories, archives, narratives) do impinge on compositional methods and forms, and do change them, because these elements rightly make strong demands; and the combination can produce new effects, insights and experiences which are very stimulating. After every such project with silent moving images, I find that my own methods and ideas of concert music composition have moved on. But, as a result of this project, I also concluded that music's own internal processes, qualities, patterns and forms can help to shape and structure the articulation of external narratives in ways that can also be illuminating, and make new connections for a range of audiences.

Notes

1. In John Stringer (ed.), *Signals* (York: University of York Music Press, 2010).
2. Sergei M. Eisenstein, *Nonindifferent Nature,* trans. Herbert Marshall (Cambridge: Cambridge University Press, 1987), 32–3.
3. James Buhler and David Neumeyer, 'Music and the Ontology of the Sound Film: The Classical Hollywood System', in *The Oxford Handbook of Film Music Studies*, ed. David Neumeyer (Oxford: Oxford University Press, 2014), 22.
4. Email 25 October 2013.
5. Peter Paret, *Imagined Battles: Reflections of War in European Art* (Chapel Hill: University of North Carolina Press, 1997), 7.
6. Danny Bright was the computer musician at the Brighton Science Festival performance.
7. John Ellis, *Seeing Things: Television in the Age of Uncertainty* (London: I. B. Tauris, 2000), 79. I am grateful to Professor David Hendy for making this connection.

Works by Ed Hughes

Battleship Potemkin (2005) (York: UYMP, 2005). fl(pic).cl.hn.tpt.perc.pf.vc.cb + tape (72') ISMN M 57020 997 2.

Strike (2007a) (York: UYMP, 2007). fl(pic).cl.hn.tpt.perc.pf.vc.cb + tape (4 channel) (88') ISMN M 57020 995 8.

Strike (2007b) Tartan DVD TVD3742/1 (2007).

Battleship Potemkin (2007c) Tartan DVD TVD3742/2 (2007).

Chamber Concerto (2010) [Composition] (York: University of York Music Press, 2010).

Dark Formations (2012). The New Music Players, Patrick Bailey (conductor), Richard Casey (piano). Recorded 28 January 2004, 26 February 2007, 28 September 2010, 8 June 2011, 21 December 2011 in London and Cambridge. Métier msv 28530, 2 compact discs.

Bibliography

Buhler, James and David Neumeyer, 'Music and the Ontology of the Sound Film: The Classical Hollywood System', in *The Oxford Handbook of Film Music Studies*, ed. David Neumeyer (Oxford: Oxford University Press, 2014), 17–43.

Chandler, David, 'Dark Formations', *Photography & Culture*, vol. 4, no. 2 (July 2011), 195–200.

Ellis, John, *Seeing Things: Television in the Age of Uncertainty* (London: I. B. Tauris, 2000).

Hughes, Ed, Katie Tearle, Sophy Rickett and David Chandler, 'Interview: David Chandler and Katie Tearle with Sophy Rickett and Ed Hughes on *AUDITORIUM* for Glyndebourne', in *AUDITORIUM* (Photoworks, 2011), 41–63.

Paret, Peter, *Imagined Battles: Reflections of War in European Art* (Chapel Hill: University of North Carolina Press, 1997).

13
To be in Dialogue with the Film: With Neil Brand and Lillian Henley at the Master Classes at Pordenone Silent Film Festival

Ann-Kristin Wallengren

The silent film festival in the little Italian north-eastern town Pordenone, only a few miles from Venice, was founded in 1982 and is regarded as the world's most prominent festival on silent film. For thirty years the week-long festival has attracted scholars, filmmakers, archive staff, musicians and film institute directors to watch, discuss and listen to silent films during an autumn week. Music has to an ever increasing extent gained interest from the visitors, and today the festival presents screenings with very varied kinds of accompaniment, not least through its master classes.

My own first visit to the festival was in 1986 when the festival focused on the pioneers of Scandinavian cinema, and the *grande finale* screened Victor Sjöström's Hollywood-produced masterpiece *The Wind* from 1928, starring Lillian Gish and Lars Hansson. All the films during the week had been accompanied by live music, mostly by pianists, but this film was accompanied by Ljubljana Radio and Television Symphony Orchestra with a newly composed score by Carl Davis who was also conducting the orchestra in the old festival cinema theatre Cinema Verdi. I was totally knocked down by the experience. The magic of silent film with live music, mentioned several times in this book, became apparent to me with a symphony orchestra reinforced by several drummers, with a score that made us feel the film's sand between our teeth and the heroine's mental breakdown in our guts. Afterwards my knees were shaky, I was rambling, I was totally filled with emotion, and I know that I behaved in a way not appropriate for a level-headed film scholar, or a doctoral student as I was at the time. In the throng after the film

I mistook Carl Davis for my husband and dragged him away, and the composer remembered me years after as 'that crazy woman'. Anyway, the magic still remains and more and more people have the privilege of discovering and enjoying silent films with live music.

The master classes in silent film accompaniment at Pordenone, organized since 2003, aim, as it says on the webpage, to 'offer new insights into film interpretation. The best film musicians, as we discover, require and develop a much deeper insight into the film's content, psychology, and structure than the rest of us, and our musicians collaborate to impart something of this, in the course of the master class sessions, in a way that is illuminating even to the most sophisticated film scholars.'[1] In this chapter I follow how these master classes work during a week in a day-by-day report, as well as explore how the art of film accompaniment in this venue has developed to be a full-fledged 'highbrow' musical art which today is elaborated in a form previously offered only to musical arts with a conventional higher status. In the chapter, I (AKW) mingle observations from the master classes in Pordenone 2013 with an interview with Neil Brand (NB) made in February 2015, and with reflections and experiences about participating as a student and musician by Lillian Henley (LH) who was one of two selected students that year (that part is written by Lillian Henley herself with a slight guidance by my questions).[2] The musicians that played for the films at the Pordenone Festival that year, as well as worked as teachers were Neil Brand, Günter Buchwald, Philip Carli, Stephen Horne, Donald Sosin and John Sweeney. The teachers lead the classes alone or together in changing constellations. The classes are always open to the audience.

Neil Brand

AKW: Neil, who came up with the idea of master classes?

NB: There was a very lovely visionary man called Jonathan Dennis who ran the New Zealand Film Festival. I travelled out to New Zealand a lot thanks to him. He died very young and was much mourned by the archive community. In Pordenone they decided to inaugurate the Jonathan Dennis lecture which would happen every year and which would involve people who weren't directly archivists but who collaborated with archivists to bring silent film to the public. I was asked to do the first one about music and I've never really talked about being a silent film pianist before. So I put together a show which was intended to be reasonably light-hearted but also as educational as possible, and to sort of tell the absolute truth about my job. And I first did that – I don't know what year the school started but I did it the year before, because the school came out of having done the Jonathan Dennis lecture – and I think it was the first time that anybody stood up in front of an ordinary audience and told them how the music worked, and what it was trying to do. And it was the first [...] attempt to look at film from the music behind [...] in terms of what pianists were looking for and what the movie was saying to us as musicians and then what we were attempting to do for an audience. And it got a lot of laughs, I also told them they were the toughest audience in the world, [...] which is true I think; ordinary audience is [...] hard to play to [...] because they are so knowledgeable and also because I think there is a kind of slightly in-built resistance to an improvising piano in some people who don't like what they consider would be a clarity of the original text, clouded by a silent film pianist [...] if you could understand.

And so two things happened from that. One was that I decided that that lecture should become a show. [...] The other was David Robinson that suggested setting up the school of music and image, and also got us a sponsor that was an extremely generous and warm man, Otto Plaschkes, who died about the same time as we begun the school of image and sound. His widow was very keen that some money from the Otto Plaschkes Trust should go to pay for the school of music and image. And the first year we set it up it was mainly me talking, but I also asked my fellow pianists if they wanted to come and talk because I didn't want it to be just my take on how you score film and music, I wanted it to be everybody's take because over the years we realized that we were all sorts of different personalities, and we all worked in such a different way. Then it was necessary to be as broad as possible, and also that that really helped in the teaching process. [...] The other thing that I insisted on from the start was that these should be public master classes, so that the students got used to playing in front of an audience. Also because I feel very strongly that I can't produce my best work without the adrenalin of having an audience sitting there while you're playing. The students I think just get used to the idea of performance even when they are struggling to find a voice for themselves or struggling to find a way in the film that there are people there taking on board what they are doing alongside the film.

Report from master class: Tuesday

(T = teacher speaking [either of the teachers], M = musician playing [either of the two master class students. They are playing almost all the time and not only when indicated in the text], F = film showing, AKW = my notes and observations from the master classes)

One teacher is on stage together with the two master class students. After a while a second teacher joins. Note that the master class students haven't seen the films before, but they play to them on first sight.

T: You can never structure an entire form, that is to say, you can never expect a structure and be able to come to an end. Often in films there are parts or frames missing, films come from different archives and look differently, and they can be edited differently.

F: The film that now starts is Alfred Hitchcock's *The Pleasure Garden* from 1925.

T: Check the opening credits, sometimes these are with still pictures. All this can give you a clue about what is coming.

F: In the film we now see typical 1920s images including young girls running down some stairs, girls on a stage dancing 1920s jazz dances. In the audience we see elderly men watching the show, and some pick up binoculars to be able to take a closer look at the girls' legs.

M: Plays jazz in a fast and dancing tempo. Slows down and plays in a more lyrical way when we see the elderly men watching.

T: Think about the beginning of the film, the exposition. In the scene it is only a dance, a show. When you slowed down when the watching men came into the picture, I think that you read in too much in the film. Nothing really happened; it was only a rather long exposition. You should not over-interpret it, but wait for obvious cues.

AKW: There is no discussion about what the pianist chose to put forward, which was obvious in the images, but on the other hand – the teacher emphasizes that this is the film's exposition, and that we have not yet got a clue about what will happen.

F: A conversation between a man and a young woman.

M: Plays in a slightly more reserved manner.

T: This is well played. This is a new sequence with a new kind of action which is demarcated and illustrated by the music.

(Another teacher enters the stage.)

T: Play with the film.

AKW: What is obvious is that both teacher and musician at this stage of the film are orientated around the most obvious and direct actions in the film, rather than for example inner thoughts of a character, or building up to something that will happen. But then again, this is playing to a film by first sight and the musicians do not know the film. This is of course a very good training, and informative also for the audience. Together with the musicians we explore the film and different ways of accompanying it.

M: Plays music that is rather blurred, that is, without any clear melody or expression. The music is rather massive though, quite fast-paced and using a wide tonal range.

T: Good. In this way you loosen and disband the music. The audience is watching the speed of the picture; usually I try to thin it out.

T: Try again; try to play only when something happens. Play much less – things in the film are not important all the time.

M: The musician tries to follow this advice, but seems to forget it after a while and goes back to the fast-paced, massive music.

T: You play as if you are forced to play. Try instead to stop and reset. Make a decision based on the rhythm on what you are feeling.

T: When the scene changes: consider if the music also has to change.

AKW: I am thinking about whether it might be useful here to have some basic lectures about film narration. Of course the teachers are very knowledgeable about this and perhaps the musicians too, but it seems that the lesson would benefit from focusing on this kind of input now.

T: Now, nothing happens for a long time so take it easy. Less is more. Is the girl in the film vicious? Make a choice, and stay with that.

T: So far we have watched the film establishing the scene milieu, now after 10 minutes we proceed to establishing characters. Get the clues but don't use them immediately.

F: The film now screened is another Hitchcock film, *Blackmail* from 1929. Opening credits, a car drives fast, children, men are arriving. A man in bed is approached by the men.

M: Plays music that sounds threatening, and then plays a bit more lyrical. The speed of the car guides the tempo the pianist chooses to apply, and she then plays softer when we see the children and the men. When we see the man in bed the music is perhaps a little overwhelming, but soon it is played slower and not that obtrusive. It seems as if the musician again has been tempted to over-interpret the action to some parts of the scene.

T: Good, you don't have to look at the people, play the scenery – light, things etcetera. Catch one thing from the film; take it down to for example a simple motif that can be used repeatedly. Sometimes make a stop and pause on some scene or story changes.

F: A long busy scene, people are sitting down at a table.

T: Relax. Let it take some time, give it space, relax. Don't force the action with music ... Here it's some tension between the man and the woman – pull out this tension even if you also play relaxation.

AKW: Now they have left a too prominent emphasis of the action, and that coincides with leaving the film's exposition. When the story develops, the musician needs to dig deeper in the narrative layers and also in the music. The mode of accompaniment becomes more complicated.

F: Again we see the table scene.

M: Now the music sounds more like a modern American musical with dissonances here and there.

T: Don't play staccato but don't use the pedal either. And slow down. We have perhaps talked too much and overwhelmed the pianist with information.

M: The pianist seems tired now and plays rather monotonous and non-melodically.

The day ends.

Neil Brand

AKW: Neil, how are the musicians chosen that are coming to the master classes? Do they apply, or do you find them yourself?

NB: It's slightly random in that all of us pianists and David Robinson particularly, have their radar out during the year. Musicians who we've heard of or who have contacted us or contacted David, who are already doing the job on a very basic level and that's the important thing. We don't want to take somebody on who has no experience whatsoever. They should have tried at least a single movie [...] who have got the itch to improvise in front of a film.

AKW: How many master class students have been in Pordenone through the years?

NB: I don't know how many alumni we've got, but it's probably heading for 25 or 30 who have been through the process and back to the big wide world and at least five or six of them are now regularly silent film musicians. And I hope they will keep coming. Because the other thing I'm very proud of is that all of us musicians have all found a way of teaching our method. I think what the Pordenone school offers is this wonderful breadth of approach. You got Gabriel Thibaudeau [...] is an absolutely unrepentant minimalist [...] he's avant-garde, he couldn't be more different in his approach than me if he tried, but that's brilliant because he would appeal to things in the music that I simply can't. Phil Carli has a very different take again, John Sweeney has a very different take again, they will all bring the best of what they know, and most of them have been playing for around 25–30 years.

Lillian Henley

AKW: Lillian, how did you become a master class participant?

LH: I contacted David Robinson, Director of the Festival, and invited him to see *The Animals and Children Took to the Streets*, a show I had composed music for (and also performed in) at the National Theatre, London in early 2013. I was really keen to gain further knowledge on how to be a better silent film accompanist and I had been admiring Neil Brand's work for a number of years. I first met Neil Brand after watching his show, *The Silent Film Pianist Speaks* in which he enlightened audiences on how an accompanying musician might interpret and score a silent film. I contacted Neil and he was encouraging of my work accompanying silent films. It was an extremely inspiring and exhausting week, but I loved every minute of it. I returned last year to Pordenone as a festival filmgoer and I was very impressed by the participants there in 2014.

I think before my time at Pordenone, my approach to silent film scoring was perhaps somewhat naive. I would spend days agonizing over a theme, studying the film for weeks before accompanying, which is certainly no bad way of scoring a film, if time was in your favour. I quickly learnt that gaining a 'screener' of the accompanied film wasn't always readily available and even if a film might be online to view, it doesn't mean the film will be exactly the same edit. It was clear I needed some quick fire improvising tools on hand. I needed to learn 'how to read a silent film' far quicker and to be armed with tools and tricks which would guarantee me a way out of any silent film sticky spot. I also realized I needed to build up my stamina. Playing an 80-minute film billing hitting every pratfall, dance number or comic routine is potentially very tiring, so I was delighted to have been accepted onto the master classes to gain inside knowledge, something I was so desperately craving.

AKW: How long have you been playing for silent films?

LH: I trained as an actor but fell into composing music for theatre and film soon after my MA in Theatre and Performance Practice, at Northumbria University. I started accompanying silent films in 2007 in London, but this was fairly infrequent and informal as the majority of my work was composing music and performing for the films with 1927 Theatre Company. My work with 1927 kept leading me to silent film, I had always delighted in the likes of Laurel and Hardy, Charlie Chaplin, Buster Keaton and George Méliès but I began to learn there were lesser-known stars or filmmakers, forgotten by history, which could equal these icons' merits, such as Colleen Moore. I fell in love with the period, the creative burst and energy of the films in the 1910s and 1920s seemed so adventurous and a breath of fresh air to watch.

My work with 1927 Theatre Company has over the years enabled me to play my own compositions live, and develop my style of musical storytelling [...] The live piano that I provide in these multi-media productions help bind the two worlds together, scoring movements from the live actors, linking it to the filmed animated characters around them, reaffirming that they exist in the same world. As an accompanist, I am often positioned between the stage screen and the audience, physically caught between the live action on stage and the audience, channelling the show's mood. I became aware of this responsibility as a 'mood effector', and I wanted to develop this language further by working with pre-existing films from the silent period. I was keen to experiment as a composer, and understand how music can vastly affect a character or the film's tone. This curiosity was crucial to my development as a composer and I was lucky to live close to the Cinema Museum in London where short silent films were often shown [...]

London is a great place to see silent films, more and more I hear of small bars or venues screening early films and I think live accompaniment is incredibly important to aid the atmosphere. Live music can unite an audience. We can feel as though we have experienced something unique, something inclusive that perhaps is experienced less when we hear a pre-recorded soundtrack in cinema. It's that live element that increases risk. The timings, tempi and melodies vary, so it can be an incredibly thrilling experience for both audience and musician. As the film is played out, the live musician has to make quick skilled choices. I like that live accompaniment to silent films can bring a certain sparkle and energy to a cinema screening. It reminds us that we have chosen to engage as a collective and for a brief moment there is solidarity in the audience, watching and listening together, taking part in something communally special.

In the months leading up to my Pordenone master class week in 2013, I had been asked to join John Sweeney and Cyrus Gabrysch (a previous Pordenone master class participant) in a new silent film night, the *Kennington Bioscope*, a regular film event at the Cinema Museum in London committed to screening rare films of the silent period. There was a light-hearted joke that audiences would 'hear the difference' after my Pordenone week, but it was true, I took no offence. I gained confidence in my playing, I learnt that I did not need to apologize for being a newcomer to the profession, and I gained a wealth of knowledge from the top silent film musicians in the world. I had never attended a class or workshop like it.

Report from master class: Wednesday

T: Try to find out about the film during the first ten minutes. The audience wants continuous thought, not slices, but you don't have to play each moment something happens. The music is a helper, but also a driver. Also try to find out about the characters: how they act and response. What different characters want you can see rather quickly, then who will try to put in some obstacle. It's useful to be familiar with how stories usually unfold, for example when it is about lovers there will always be a conflict: try to musically anticipate this conflict.

T: You don't have to play the rhythm all the time; you can have the rhythm in your head and let it glimpse through here and there. To play the tempo or the rhythm all the time can force you into a play that you don't want to have and that you don't think is good for the film. Try to find structures by only using some chords. The most beautiful chord is major 7 which carries a lot of possibilities. A trick is to play that chord some octaves higher. That could also be useful when you don't know what is going to happen. You must be in dialogue with the film.

T: We feel time and rhythm before we feel mood. The rhythm, and what we do with it, is very important. Try to bring the music to an end. Major and minor keys are extremely powerful tools and change mood and narrative directly.

The keys can also very fast catch and express complex feelings that are simultaneously present. The goal is to not think in a musicological way, in keys, harmonies, etcetera, but to just processing.

F: A film with Charley Bowers will be screened as the teacher recommends playing to comedies as a practice, since comedies from this time often have a very clear-cut and plain dramaturgy. The teacher himself is playing to the film to demonstrate how music can influence the narrative and comical impact. It is the only time during the master classes that the teacher is playing more than some bars. The music makes the film much more funny, it is a remarkable difference.

T: In these kinds of films you need musical help and support to build up the jokes. There are not gags all the time, sometimes the music must rest. In comedies you can influence the speed; you can increase or decrease the speed very much with music.

(Another teacher enters.)

T: Be very careful with minor keys in comedies since minor keys endow a depth with the film which is probably not present in the film... You find the REVEAL, and this is very important; now it's time for the big epic theme.

F: The teacher plays to the end of the film to illustrate what he has said. The accompaniment makes the film hilariously funny.

AKW: The silent film pianists who are teachers in the master classes can in some way be regarded as the heirs to the authors and musicians during the silent era, who wrote classical books like Erdmann and Becce's *Allgemeines Handbuch der Film-Musik* and Rapée's *Encyclopedia of Music for Pictures* and the like.[3] Some things pronounced in the classes are recognizable from these manuals. A crucial difference is of course the master class situation, the direct and live teaching in the art of accompanying silent films, and of course also the totally different and changed historical situation.

(The discussion now leaves the comedy genre and proceeds to how to play to documentaries, and one of the master class students is now playing.)

T: Documentaries very seldom embody a natural structure, a natural narrative. One purpose with music to film is storytelling, so in documentaries and actualities: try to find the story. It is up to you to bring out the impact.
T: In all types of films it can be useful to now and then use a little mickey-mousing and to accentuate important actions with a sforzando, like in the Charley Bower scene when the egg is smashed into pieces or when the wall

falls. If we watch Lumière's 'Workers Leaving the Factory' it is important, in getting context, to think of the people who made the film at the time. In documentary we must take an interest in the people then, it is there we find the story. Think of the people leaving the factory, they have been working the whole day, they are sweaty and now relieved that work is over. Documentary is the hardest thing to play.

AKW: I am thinking that the question about playing to different genres even more entails the necessity for knowledge about film history and film narration.

T: Find what it is that moves you and play that.

The day ends.

Neil Brand

AKW: Neil, is it possible to teach how to play to silent films?

NB: I can't teach people to improvise on the piano – they can either do that or they can't – I mean it's a very hard thing I think to learn. I think it's a hard thing to teach and what you are looking for is people who have an instinct to be able to just improvise music that fits to an idea.

AKW: You tried in the master class to guide the pianists to find the inner idea of the film.

NB: Yeah, I think reading the film is the one thing you can teach, what the film is, what the genre and the style is, we can give them something to look out for. But actually playing a movie you've not seen, I think draws on the most creative element of you. And you will come up with your best music, I believe, initially, when you are in front of an audience and finding your way through a film.

AKW: So you mean the audience plays a role in this?

NB: A huge role, yeah. Without any doubt at all. It's harder in Pordenone when you're right there in the pit, and you can't *see* them. You can hear them. But under normal circumstances when you're sitting at the piano in the main body of the hall, looking up at the screen when everybody else is looking at the screen, the audience is palpable. It's not just any vocal reaction you can note on the film, but the silence, pregnant silence of the audience. You can feel that too.

AKW: You feel the atmosphere in some way.

NB: Yeah, I try to judge my audience in the first 20 minutes, as to how much work I need to do to make this film come across, if at all. What I'm looking for is how far they react as one entity which is always what you're looking for eventually. How far there are people laughing at the film, that you have to try help the film steamroll through moments that might allow that. If it's a comedy how much they are laughing.

Lillian Henley

AKW: Lillian, describe how it is to play to a film which you watch for the first time?

LH: Neil Brand explained the exposition of a film and spoke about ways to examine film techniques. He explained you could learn a lot about a character by looking at the camera frame and the length of time the filmmaker has spent on a character. He also showed us that if the filmmaker creates a 'reveal' the accompanist can enjoy this moment. It's a gift and a chance to play how a plot unfolds.

Neil showed us the opening of *Lazy Bones* (1925, Dir. Frank Borzage) and he explained how we could make educated guesses about the characters in the film. The lead character, where the film gets its name, first appears sleeping in a hammock for a considerable length of time, in full shot. We start to imagine the comedic possibilities and even dramatic tensions that this character might experience as the frame cuts to a pretty girl arriving at his family home, and introduces older members of the household. I have since learnt to enjoy this guessing game, trying to be one step ahead, or at least pretending to be ahead of the game, it can be incredibly rewarding. Of course, any decision we make as silent film pianists can go wrong, so if you can guess the majority of the film right the audience should forgive any moments you might miss.

There is only so much you can prepare as a silent film pianist. We are lucky to have access to a lot of films online, so as much information an accompanist can obtain is essential. Trying to see as many silent films as possible and obtaining a wider knowledge of music from the period is a good idea. But if you have not seen the film, there are often surprises you can never predict. For example, when a film gives a close up of a piece of music or title of a record that may have been popular at the time, play something close to that mood and play it confidently. Stephen Horne showed us that you just have to keep going, even if you have never heard of the dance track the couple are suddenly dancing to, it is about how you deal with these surprise moments that keep the film accompaniment interesting.

Report from master class: Thursday

The day starts with some meandering discussions about different films and how to play to them. First there is a kind of experimental documentary film showing flowers that blossom in speeded-up motion. The teacher speaks about different solutions to these pictures, how it is possible to characterize different flowers, and the advice is to go 'into the texture of the image'. After that the work concentrates around a documentary from the First World War, with pictures from the trench, and the teacher asks the students how they reacted when the saw the pictures: Were they thinking of war sounds? What would they like to play? The pianists have different solutions but seem a little uncertain, and the teacher then shows the clip with the music recommended when it was shown at the Imperial War Museum. It is a funny type of cavalry tune which sounds inappropriate, not least to a very sad scene where a young man dies. A Danish scholar in the audience comments that when the film was shown in Denmark there was musical silence to that particular scene. Then a scene from a British feature film is shown to which the accompanying music is recorded and a short discussion about different ways to accompany the scene follows. A second teacher joins the party and another phase of the lesson follows.

T: Try to reduce the film to a single motif which can be used at times when you don't really understand what's happening.

F: A scene from the British film *Piccadilly* from 1929 with people dancing.

T: Somewhere in the dance by some of the characters you find the tempo that you should try to bring forth. You shall not play only 'dance' but the feeling/action that you find by one of the characters and which make the dance necessary.

AKW: This seems to be a good advice which is very close to the film narrative.

(The teacher now gives instructions about how to play to dances: between 1923 and 1924 the dance was often a waltz, after that more modern dances came, like charleston.)

T: We see here: an establishing shot – the streets of London – a place of entertainment – a restroom. That is, less and less space, analytical editing, a kind of zooming. Try to find that grading also in the music, play the big in the small. For example, when it's dialogue you must play softer.

M: The student plays charleston in a fast tempo but misses the dialogue, which caused the teacher to comment on that.

T: When you test some music which you are not sure will be fitting, try it for 15 to 20 seconds: if it's not appropriate you must then change. Never longer than 15 to 20 seconds. Of course, there are no such 'rules' when the music fits the scene, but think about varying the motif.

(The teacher goes on pondering about films with dance. He thinks that they used dances during the production of the films also as a kind of instruction to the musician, a guidance to what to play and in what tempo. He thinks that it's one of reasons why the film is cut back and forth to the dance scenes: to know the tempo and the type.)

M: Now the student plays rather loud and pompous, using a broad range of the piano. This is not the best way to play to in that particular scene, the teacher thinks.

T: If you go huge the audience expects something very important is going to happen. Be cautious and save that kind of play until it's needed.

AKW: The students have made a considerable improvement during the week. It is a very different experience now to listen to how they choose to accompany the film scenes. The kind of teaching during the week has also changed, from more abstract discussions to very tangible and instructive advice. It seems to be effective pedagogy working this way.

(For the rest of the film scenes the teacher talks very much about finding who is the protagonist, what kind of person he/she is, what relations are perceptible, how the characters develop, and so on.)

The day ends.

Neil Brand

AKW: Neil, the students have a kind of examination at the end of the week, on Saturday evening?

NB: Yes, another element of the school that was really important when we started the master classes is that at the end of the week the students get to play a movie in full in front of the audience [...]

AKW: When they are playing to the full film, do you notice if they also have the ability to interact with the audience which we talked about earlier?

NB: I think they can *hear* an audience reaction, and it does work on them. But to be frank, they're so concentrated on the film. Two days or so in beforehand, they get freaked by the idea of having to sit and play a film they don't know of. And particularly in front of the most distinguished silent film scholars in the world [...] I try and listen out for every last little message the audience is giving me when I'm playing. It's very useful, very interesting to hear. And you can, I think you can skew your music so that a film is being given much more support to hit an audience when it's not working so well. Or you play [...] on soft pedal and really allow a film to do its work only just give it enough music to keep it slightly airborne. Those are the films that are a joy to play, really, you are there only to facilitating, you are not there to do anything more than that.

AKW: So you enjoy that much more as a musician? When you keep them airborne as you say?

NB: Yeah, I don't want to work so hard. I don't like the idea that a movie that I know works, isn't working because that audience isn't allowing it. That's when I feel the music has to step in, take a hand. This is not about comedies not getting laughs, this is about dramas not hitting home because an audience is finding it hard to get their heads into the ... the attitude, or whatever, that's when it becomes hard work, when you have to sell a movie to an audience. That is also something we teach them. And I think once students have done three or four performances, they can feel that something is working. It still amazes me that there are people out there who don't understand comedy, who play comedies and don't get any laughs and they don't understand that they are not playing those comedies very well. Astonishes me, really. There we are. There is an element of collaboration that has to be assumed from the start in musicians that play for silent film. That's what makes the film work, not shining the spotlight on their own ability.

AKW: Are you also inspired by sound films?

NB: Oh God, yes. I've recently done a BBC series. That's always been my main interest [...] film music as [...] sound film. And I've always applied that to silent film because I think that's the idiom people understand. If you are going to play a horror movie from the silent era you use the genres and the tropes that people expect from horror movies today. So you go in with that intention in mind. I try [...] period movie [...] I don't intend to go massively modern; I certainly don't use jazz or blues *unless* the film demands it. But I try not to carry it away too far from the 1920s or the 1930s by the latest, in terms of the music. But if they are going to watch a Western, then make it sound like a Western, give it that Elmer Bernstein sound. Give it that Bernard Herrmann sound for thrillers, you know. That's exactly what people sort of respond to because that's what they know. I don't teach that perhaps, and that's not [...] I don't impose that on other people. That's my way in. If that's the way they want to go with it – fine – but the problem is that going that way doesn't necessarily allow you to have your own voice. But I think the main thing we're teaching is trying to get the pianist finding the voice that is theirs, not the voice that belongs to somebody else [...] And so the students will go away with an extraordinary experience [...] getting on with the job themselves.

Lillian Henley

AKW: Lillian, what are your thoughts about the accompanist's role to silent films?

LH: Yes, one question that arose from my week at Pordenone was exactly that: what is the role of the accompanist? Is it showmanship, is it subservience, or somewhere in between? In most part, all the silent film accompanists were in agreement with each other, that the accompanist should not play a subservient role. I think the role of the accompanist is to serve the film sensitively, whilst putting your own interpretation onto the film. I quite like thinking of the film as if it was a solo singer, and as a pianist would accompany the melody of the soloist, they work together, with the pianist supporting, helping showcase what the singer can do and subsequently the pianist. If the singer wants to slow down a section, it's up to the live accompanist to bend its time signature to the soloist, playing a supportive role but always playing competently and stylishly. In film accompaniment I want the music to enhance the spectacle of the film. I want audiences to remember the emotional journey of a film, whilst of course, still taking advantage of any opportunity when the film can allow for the accompanist's musical agility. I do not play subserviently but I do feel my role, as an improvising pianist, is to brighten and colour a film with a sensitive touch.

AKW: What do you think about the importance of music to silent films?

LH: I am always struck by the responsibility I have as a composer, to play something that works with the film. Some friends of mine specifically go to see a silent film with the intention of hearing a particular pianist. That is always a lovely compliment for that pianist. Since Pordenone,

I have been keen to develop my own style of accompanying as well as taking inspiration from my colleagues, so hopefully audiences enjoy my slant on the films I accompany.

My work as a composer has always started with piano improvisations, which I then set into a piece of work. Improvising on the piano has always come naturally to me. It's a kind of stream of consciousness, which I enjoy pouring out any thoughts or feelings into. I often use a still image, or photo, a painting, to create a musical theme as my starting point in other areas of my work. However, when I am improvising to film, the images are constantly moving and evolving, the possibilities of the plot are endless and there are streams of characters you need to get to know very quickly. Not only that, you have to make decisions about characters, sometimes in a single shot, in order to help cement the mood or plot for the audience. You learn when to read if something is significant and whether to underline this in your live score, and you hope you haven't underlined any red herrings too heavily!

Report from master class: Friday

Today is the last day of master classes and the students are going to work a little on the film that they are going to play to in full the following evening, where they will play half the film each. The film they will be playing is the Swedish drama *Rågens rike* (*The Kingdom of Rye*, 1929, Ivar Johansson). But before going into this work, they test themselves through playing to a totally different genre: a British whodunnit movie. Two teachers start out this morning.

T: The choices are more or less endless when a film starts, but it narrows by and by.

M: The film starts and one of the students is playing, and chooses to play in a charleston style. For example, at the beginning of the film someone runs and the student plays in a very dramatic way. After that there is a scene with people joining together for tea party and the music played is very joyous, almost like it is a big party. The teachers seem to think that this charleston style is a little bit over the top, and later during the day advise how to think economically about music, probably based on this experience. On the other hand it is not surprising that the student chooses to play charleston, considering the lesson yesterday which in one part focused on dance music.

T: Don't be too lavish with the charleston music. There is a lot of tea and small parties in the 1920s cinema, so if you choose charleston as a basic motif for such scenes, you will have to play that very often. Take it down a little, only have some jump in the music.

T: Here you should build up an armoury of understandings and it's important to use it, and then it's possible to go ahead in the music. If you succeed in your interpretation you can be ahead of the actions and then it's very effective.

F/M: Nothing in particular seems to be happening in the film and the students are having a hard time trying to find an accompaniment that takes out some element from the narrative. The searching quality in the music is very obvious.

T: You sometimes find that the film is uneventful and monotonous and that you play this, that and the other. This is an occasion when a film needs a driver through the music. As here, we have a murder, and it will shortly be followed by a quest, which often is quite boring or lacking in action in a silent film. The energy behind the quest can be music which injects energy, but it can also be dangerous if played wrongly. You can play louder, and it's often a question about rhythm, or you can play a motif that sounds 'mysterious', but you must try to give a sense that we are going somewhere, music must be a driver.

(The teacher himself demonstrates at the piano playing to scenes that really are very boring – only men talking, no movements, only a flashback disrupts the totally uneventful sequence.)

T: There will appear many perspectives, and how do you translate this into music? Try to find a motif for each character, at least the important ones.

(Two more teachers enter the stage; together they are now four teachers. They start to work with the Saturday evening film *The Kingdom of Rye*, but the film is not shown yet, and they only talk about the kind of rural landscape that the title suggests.)

T: Play the landscape in the film without starting the film. Take in a theme from the landscape painting and try to keep this theme during the film.

T: Bring down the music to a smaller scale, to characters, to different perspectives but the same kind of space.
(A fifth teacher joins the party.)

T: Play the same thing again but make it smaller. Imagine that you don't have a piano but only a flute and a clarinet, that you only can play two tones at a time in two single lines. Think minimally as a test. And use only whole notes and half-notes, play really slow, and then we can hear the characters.

AKW: This turned out to be a very interesting instruction which led to a totally different, but very rewarding way of playing the film since it seemed to be closer to the film, to be an organic part of it.

T: Imagine different instruments, it gives a better psychological impact if you play slowly and simply ... If the emotion in the film is strong, you don't have to do much. Choose your moments otherwise the audience gets tired and lost in the massiveness of the music. You must give them pause for breath. Sometimes use only one hand, don't think only in clouds but in lines. A cheap trick is to change register but not necessarily with both hands. This will change the colour. You can internally imagine a song or the sound of an instrument; this will change how much you think you should play.

(Now the film starts.)

T: Stay away from big musical statements in the beginning, unless it's a big fight or something. If you play too big – you have nowhere to go. If you don't know what to do, play ostinato – a tone or arpeggio.

(Musicians in the audience now interfere in the discussion and suggest that the students could play a simple motif, a simple base, something that puts forward a 'Scandinavian' mood. A lot of things are happening in the film and the student tries to adapt to all this but gets the advice to avoid trying to musically catch everything in the story, but rather play more 'in general'. Everybody agrees about playing too much is a very general problem among most silent film musicians.)

T: Imagine that you only have one thousand notes to play – you have to economize. Or that every tone costs you five dollars.

T: In Scandinavian films you often find long scenes when not much is happening, but they give colour, narrate about locality, and so on. Think these scenes as accompanied by someone who sings with local colouring. Again, play the scenery – here it's very neat with the wetnurse, the barn. It's empty, spare, clear-cut. Play the look, the composition of the picture.

The day ends.

Lillian Henley

AKW: Lillian, tell us about your experience being a master class student at the Pordenone Silent Film Festival.

LH: Over five mornings, Anne-Maartje Lemereis (my fellow master class participant) and I would take it in turns to play for our silent film tutors, and this came with highs and lows; nervousness, excitement, frustration (at one's own fingers) and triumphs. I felt fraudulent (I hadn't been to music school), and I was beginning to feel that perhaps I was here by mistake, I soon realized I just needed to relax so I could play the films with a sense of fun, which is why I love to score films.

The master class experience taught me to create music that I would enjoy listening to, and to always be in control of the melody I was improvising. Günter Buchwald particularly highlighted that the accompanist should always play 'great music'. As musicians, we should complement the film accurately, but the music foremost must be of high quality. The accompanist should develop strong musical themes and for the theme to be determined by the film's pacing or rhythm. In one exercise, Günter asked Anne-Maartje and I to think of two themes for the film *Beggars of Life* (1928, William A. Wellman), one for the train and one for the lead characters on the train. He shared his motif he had created for the train (integral to this film) and explained how important it is to incorporate musical themes over the top of this, building the film's sound around that musical idea. Anne-Maartje and I both improvised with Günter, whilst he played his violin, and I was struck with how keenly he listened to our music. This of course meant my nerves increased and my music worsened! I realized then, I had previously hidden behind the film and I needed to present my music score with more awareness. Now I like to start each film's theme with a better sense of my own presence and with the knowledge that my accompaniment is worth listening to.

Neil Brand also confirmed that the player should feel at ease, when developing themes for films. Neil complimented me on my musical ideas, but I should strive to develop the themes further before changing to another musical motif. The live music should measure up to the film you are scoring and sticking with a theme cements the ideas you are sharing with the audience. I would often chop and change my ideas for a film far too readily, partly because I was changing my mind as I played. I often think about this before I play a silent film and try to hum a melody to myself before a film starts so I play more decisively.

John Sweeney introduced the notion of the thread of a film, to envisage how a picture can also be a melodic line. To take this literally, one can play a one finger melody, it might not always be necessary to play ten fingers all at the same time. Donald Sosin echoed this, and encouraged me to imagine the piano score as an orchestra. What might be the oboe solo be, or the first violin section? Or where's the string section, the horn section and at what point might a full orchestra join in? The piano can mimic so many different instruments, it's a wonderful versatile instrument to play, and I found it exhilarating to remember its versatility.

As an exercise, Donald asked us to play *The Hunchback of Notre Dame* (1923, Wallace Worsley) with Lon Chaney starring. I had enjoyed seeing this a few years ago, with John Sweeney brilliantly accompanying in London at The Prince Charles Cinema. The street scenes were epic, full with large crowds and I had originally thought I would struggle with the immensity of the picture. I attempted the film in the class and I took the audience to the 'wonderful streets of Seville in Spain' not fourteenth-century Paris. In my keenness to develop a busy city motif, I had chosen a musical mode that was closer to traditional Spanish music rather than a Parisian sounding theme. Gaining a broad repertoire became suddenly top of my to-do list. I understood that choosing certain kinds of music ideas that are indicative of a place supports the film, and signposts the period further. Adhering to the obvious is not necessarily playing a musical cliché. I only like to play the 'obvious' for a short period of time, for I would rather play a character's emotions, which I feel is always a complex and interesting challenge. Sometimes the filmmaker gives you, however, no choice, if you're in an exotic place and there is nothing else on screen, you should play the sound of that place.

An old habit that I loved to do was alternating between major and minor chords but as Philip Carli pointed out, although it was a nice technique, the current scene being played on film, was 'just children playing'. The filmmaker in Philip's master class session had not shown any signs of things feeling disturbing or sinister, it was a pleasant tranquil setting. It was clear I needed to disguise my 'thinking' music better so not to mislead the audience. Ostinato, a repeated musical idea or note, was a favourite of the tutors if you were waiting for something to happen, or taking a moment to decide a mood, and it's been particularly useful to remember that.

AKW: The last evening the master class students always play a full film in a regular screening for the Pordenone audience, as a kind of 'examination' you could say. How did you experience this?

LH: Anne-Maartje and I shared our film debut at Pordenone, Le Giornate Del Cinema Muto. We accompanied *Rågens rike/The Kingdom of Rye*, a very long Swedish film that was shot beautifully. It's very rare for any accompanist to share a film in two separate halves, but we were ready for the challenge. I played second and I remembered trying to take on the musical motifs Anne-Maartje had established so as to provide continuity for the audience. I think we both felt nervous, but we were encouraged by our tutors. I remembered playing some dissonant staccato chords in a confrontation scene between the villagers. The image was so stark, angled with extreme angered faces. I felt that after I had played loudly, I could take a moment to pause. (A technique I remembered Günter being a fan of.) The images of these people were so powerful I felt I could come off the piano until the frame cut to another shot. Suddenly, a large fight broke out on screen, I was pleased I had made that decision, so it felt like an effective choice to wait, my fingers poised for the full blown action that was about to commence.

I learnt after my performance that you cannot hurry a film. If the film is slow, go with that, there is no point trying to play something that is not on the moving image. Philip Carli reminded me of this: if the filmmaker wants to linger on a shot, it's a good idea to stay there musically too, it's important to present a musical vision that complements and completes the filmmaker's vision too, there is no point rushing, and it can be distracting for the audience.

Neil Brand

AKW: Neil, one point of departure for this book is the importance of music for the popularity of silent film today; that music is a way to increase the interest in silent film. Do you agree?

NB: Yes, very much so. I think there was initially a slightly unhealthful approach to silent film in that musicians who already had a following would play to movies without necessarily understanding the movies, in producing music that hadn't any link to the film. I went to a couple of events where basically the music was the star and the film was playing, almost irrelevantly, in the background. 1990s into the 2000s. That's really not helping much at all because the film is not allowed to speak. But that has changed radically and I think it's partially a new generation of musicians and silent film enthusiasts who discovered silent film either through universities or just through DVD, started going to live shows, began to realize what kind of [...] what the magic is between film and music. And that is now I think making a huge difference. People who are undecided about going to see a silent film they don't necessarily know, will make up their mind if it's also a musician they like. And that's making a big difference. I've always felt that the point of the musician was not kind of concreting the film into the period, with period music, but to form a bridge between the film and the modern audience, so the two come closer together. So the music is speaking to a modern audience in ways they understand, even if the film isn't. That's what makes the silent film work with the modern audience [...] And that's always the thing with silent film, we're not trying to say these are retro masterpieces, we're trying to say, these work now.

Notes

1. http://www.cinetecadelfriuli.org/gcm/giornate/masterclasses.html, accessed 15 February 2015.
2. The interview with Neil Brand was conducted via Skype, and the interview and discussions with Lillian Henley was conducted via email. I am very grateful to both that they wanted to contribute to this book, and for the time they offered to interviews and other correspondence.

   Neil Brand has been accompanying silent films for nearly 30 years, regularly in London at the Barbican and NFT, throughout the UK and at film festivals and special events around the world, including Australia, New Zealand (three times), America, Israel, Sweden, Denmark, France, Germany, Ireland, Luxembourg, and, in Italy, the Bologna, Aosta, Bergamo and Pordenone festivals where he has inaugurated the School of Music and Image. Training originally as an actor, he has made his name as a writer/performer/composer, scoring BFI video releases of such films as *South* (Shackleton's journey to the South Pole), *The Ring* by Alfred Hitchcock, the great lost film *The Life and Times of David Lloyd George* and early cinema, avant-garde cinema and Russian pre-Soviet cinema; and new DVD releases for the Danish Film Institute and Lobster Films, Paris as well as a highly acclaimed jazz score for the 1927 Anna Mae Wong film *Piccadilly* (BFI video) which premièred live in September 2003 at the Lincoln Centre, New York and the Barbican concert hall in March 2004. His most recent DVD scores are for *Silent Dickens, Silent Britain, The Cat and the Canary* (1927) and the Laurel and Hardy short *You're Darn Tootin'*, commissioned by Paul Merton with sound effects provided by the audience. From webpage http://www.neilbrand.com/music_biography.shtml, accessed 30 April 2015.

   Lillian Henley is a composer, silent film pianist and performer. Lillian is also an associate artist of 1927 Theatre Company. She tours and performs with 1927 Theatre Company and has been responsible for the live music since 2006. Awards with 1927 Theatre Company have included Herald Angel, Fringe First, Carol Tambor, Arches Brick, Total Theatre, Off West End, Peter Brook Empty Space, and Critics Circle Awards. Lillian Henley enjoys composing for film, and regularly plays at Kennington Bioscope at the Cinema Museum, London, with Cyrus Gabrysch and John Sweeney. Silent Film Performances have also included The British Silent Film Festival (2013/14), inHouse Film Festival, London (2014), Fashion in Film Festival (Barbican, 2010) and Laurel & Hardy Double Bill (Hoxton Hall). Henley was a master class participant at the Le Giornate del Cinema Muto, Pordenone, Italy in 2013 and continues to strive to create quality film music. She is currently producing an album of music from Golem 1927's recent theatre show and is developing her solo piano album 298.

3. Hans Erdmann and Giuseppe Becce, *Allgemeines Handbuch der Film-Musik* (Berlin: Schlesinger'sche Buch u. Musikhandling, 1927); Ernö Rapée, *Encyclopedia of Music for Pictures* (New York: Belwin Inc., 1925).

Bibliography

Erdmann, Hans and Giuseppe Becce, *Allgemeines Handbuchder Film-Musik* (Berlin: Schlesinger'sche Buchu. Musikhandling, 1927).

Rapée, Ernö, *Encyclopedia of Music for Pictures* (New York: Belwin Inc., 1925).

Index

Note: Numbers in *italics* refer to illustrations.

CPSIA information can be obtained
at www.ICGtesting.com
Printed in the USA
LVOW04*2235190416
484424LV00013B/77/P